QUANTITATIVE METHODS FOR SURVEYORS

QUANTITATIVE METHODS FOR SURVEYORS

CARL ROBINSON & LESLIE RUDDOCK

CONSTRUCTION PRESS London and New York

Construction Press
an imprint of:
Longman Group Limited
Longman House, Burnt Mill, Harlow
Essex CM20 2JE, England
Associated companies throughout the world

*Published in the United States of America
by Longman Inc., New York*

© C. Robinson and L. Ruddock 1984

All rights reserved. No part of this publication may be reproduced, stored in a retrieval system, or transmitted in any form or by any means, electronic, mechanical, photocopying, recording, or otherwise, without the prior permission of the Copyright owner.

First published 1984

British Library Cataloguing in Publication Data
Robinson, Carl
   Quantitative methods for surveyors.
   1. Surveying
   I. Title
   II. Ruddock, Leslie
   526.9   TA545
   ISBN 0-86095-891-4

Library of Congress Cataloging in Publication Data
Robinson, Carl, 1942–
   Quantitative methods for surveyors.

   Includes index.
   1. Surveying. I. Ruddock, Leslie, 1950–
II. Title.
TA545.R645   1983   526.9   82-19710
ISBN 0-86095-891-4

Printed in Singapore by
Selector Printing Co (Pte) Ltd

TO MARGARET & PAULINE

CONTENTS

Preface		xiii
Acknowledgements		xv

I LAND SURVEYING 1

1	*Land surveying procedures*	2
1.1	Introduction	2
1.2	The classes of survey work	2
1.3	The principles of surveying	3
1.4	Survey accuracy and control	4
1.5	Survey and plotting procedures	6
1.6	Exercise	7
2	*Map projections and the Ordnance Survey*	8
2.1	Selection of a map projection	8
2.2	Map projection systems	9
2.3	The National Grid	13
2.4	Sheet numbering and the National Grid	16
2.5	Maps produced by the Ordnance Survey	16
2.6	Parcel numbers	20
2.7	The accuracy of Ordnance Survey maps	21
2.8	Exercises	23
3	*The chain survey*	24
3.1	Introduction	24
3.2	Survey principles	24
3.3	Survey equipment	26
3.4	Survey procedure	28
3.5	Chaining a line	29
3.6	The location of detail	30

vii

Contents

3.7	Setting out right angles	31
3.8	The fieldbook	33
3.9	Correction for sloping ground	35
3.10	Obstacles to field measurement	37
3.11	Accuracy of the chain survey	39
3.12	Plotting the survey	41
3.13	Exercises	42
4	***Bearings***	**43**
4.1	Types of bearing	43
4.2	Choice of meridian	45
4.3	The prismatic compass	46
4.4	The compass traverse	48
4.5	Exercises	51
5	***The theodolite***	**53**
5.1	Function of the theodolite	53
5.2	The principles of the instrument	53
5.3	The axes of the instrument	55
5.4	The components of the instrument	55
5.5	The simple telescope	56
5.6	Types of theodolite	58
5.7	Specification of the theodolite	61
5.8	Setting up the instrument	62
5.9	The measurement of horizontal angles	66
5.10	The booking of observations	69
5.11	The measurement of vertical angles	70
5.12	Sources of error	71
5.13	Exercises	74
6	***The traverse survey***	**75**
6.1	Accuracy of the traverse	75
6.2	Forms of traverse	76
6.3	Field procedures	77
6.4	Advantages of the theodolite traverse	78
6.5	Electro-magnetic distance measurement	78
6.6	Traverse accuracy and plotting	79
6.7	Example of a theodolite traverse calculation	86
6.8	Plotting the traverse	94
6.9	Methods of reducing the calculations	95
6.10	The traverse table	99
6.11	Exercises	100
7	***The level***	**102**
7.1	The principles of levelling	102
7.2	Components of the level	103
7.3	Types of level	104
7.4	Accuracy of levels	107

7.5	The specification of the level	109
7.6	Use of the level	109
7.7	The levelling staff	110
7.8	Permanent adjustments to the level	112
7.9	Exercises	115
8	*Levelling*	117
8.1	Introduction	117
8.2	Bench marks	118
8.3	Levelling procedures	119
8.4	Booking the levels	121
8.5	Inverted staff readings	125
8.6	Sources of error when levelling	126
8.7	Applications of levelling	129
8.8	Exercises	133
9	*Tacheometry*	135
9.1	Advantages and limitations	135
9.2	Vertical staff tacheometry	136
9.3	Horizontal staff tacheometry	155
9.4	Exercises	158
10	*Aerial photography*	160
10.1	Introduction	160
10.2	The principles of aerial photogrammetry	160
10.3	Conditions required for a true plan	161
10.4	Equipment	161
10.5	The principles of vertical photography	161
10.6	Distortions of aerial photographs	163
10.7	Stereoscopy	167
10.8	To baseline a stereopair of photographs	168
10.9	Radial line plotting	168
10.10	Determination of ground height	169
10.11	Mosaics	170
10.12	The characteristics of photo images	170
10.13	Exercises	172
11	*Areas and volumes*	173
11.1	Introduction to areas	173
11.2	Area calculation procedures	173
11.3	The planimeter	181
11.4	Calculating areas from rectangular co-ordinates	183
11.5	Volume calculations	186
11.6	Cross-sections	187
11.7	Volume of mean areas	188
11.8	Prismoidal formulae	189
11.9	Cross-sections combining cut and fill sections	191
11.10	Volume from contours	193

Contents

11.11	Volume from spot height	194
11.12	Exercises	196
12	*Setting out*	197
12.1	Introduction	197
12.2	Setting out earthworks	198
12.3	Setting out buildings	201
12.4	The vertical control of a structure	208
12.5	Exercises	211
13	*The measurement of existing buildings*	212
13.1	Equipment	212
13.2	General procedure	212
13.3	Booking procedures	213
13.4	Rectified photogrammetry	216
13.5	Plotting	217
13.6	Exercises	218

II STATISTICAL ANALYSIS 221

	Introduction	221
14	*Data and its presentation*	222
14.1	Introduction	222
14.2	Organising data: frequency tables	222
14.3	Grouping of data	223
14.4	Cumulative frequency	226
14.5	Graphical presentations of data	228
14.6	Other methods of displaying data	234
14.7	Exercises	239
15	*Measures of location*	240
15.1	Introduction	240
15.2	Measures of central tendency	240
15.3	The arithmetic mean	240
15.4	The median	244
15.5	The mode	247
15.6	Other averages	248
15.7	A comparison of the mean, median and mode	249
15.8	Quartiles, deciles and percentiles	249
15.9	Exercises	250
16	*Measures of dispersion*	252
16.1	Introduction	252
16.2	The range	253
16.3	The quartile deviation	253
16.4	The mean deviation	253
16.5	The standard deviation	255
16.6	The coefficient of variation	259

16.7	A note on the interpretation of the standard deviation	259
16.8	Skewness	260
16.9	Exercises	262
17	*Probability and the normal distribution*	264
17.1	Probability	264
17.2	The rules of probability	265
17.3	Permutations and combinations	267
17.4	The binomial distribution	268
17.5	The normal distribution	269
17.6	The normal curve	269
17.7	The significance of the normal distribution	271
17.8	The standard normal distribution	271
17.9	Z values and areas under the normal curve	273
17.10	Areas under the normal curve and probabilities	275
17.11	Exercises	277
18	*Index numbers*	279
18.1	Introduction	279
18.2	Single-item index numbers	279
18.3	Multi-item index numbers	282
18.4	Comparison of Laspeyres and Paasche indexes	283
18.5	Price relative index	284
18.6	Choice of base year	285
18.7	Chain-base index numbers	286
18.8	Use of index numbers	286
18.9	Exercises	287
19	*Linear regression*	289
19.1	Introduction	289
19.2	A linear relationship	289
19.3	The line of best fit	292
19.4	The method of least squares	292
19.5	Use of the regression line for prediction purposes	294
19.6	Exercises	295
20	*Correlation*	297
20.1	Introduction	297
20.2	Scatter diagrams	297
20.3	The product moment correlation coefficient	299
20.4	Example of positive correlation	301
20.5	Example of negative correlation	302
20.6	Rank correlation	303
20.7	Spearman's coefficient of rank correlation	304
20.8	The interpretation of the value of r	306
20.9	The correlation coefficient formula: a mathematical note	306
20.10	Exercises	307

Contents

21	*Time series analysis*	310
21.1	Introduction	310
21.2	The method of moving averages	311
21.3	Cyclical variations	312
21.4	The use of time series analysis	315
21.5	The semi-average method	317
21.6	Time series and the least squares method	317
21.7	Exercises	320
22	*Sampling methods and sampling distributions*	322
22.1	Introduction	322
22.2	Sampling methods	322
22.3	Sampling distributions	324
22.4	The use of confidence limits in the estimation of means	326
22.5	Significance tests	327
22.6	Small samples	329
22.7	Estimation of proportions	330
22.8	Exercises	330

III DATA PROCESSING SYSTEMS 333

23	*Computer systems*	334
23.1	Introduction	334
23.2	The components of a computer system	334
23.3	Computer programs	336
23.4	Flowcharts	337
23.5	The role of the computer within the information system	338
23.6	The location of the computer within the information system	339
23.7	The benefits of automatic data systems	344
23.8	Exercises	345
	Appendix The normal distribution function	346
	Answers to selected exercises	349
	Index	352

PREFACE

The modern surveyor must be capable of appraising information and formulating decisions concerning an ever increasing range of disciplines allied to his profession if he wishes to benefit his practice. For this data to be realistically analysed, it is important that every new surveyor entering the profession obtains an understanding of how survey observations can be recorded and the way that the subsequent results may be presented, so as to have maximum significance to others.

This book has been designed to satisfy these demands and reflects the requirements of the Royal Institution of Chartered Surveyors (RICS) Part I examination in Quantitative Studies. By outlining the principles of measurement, statistical analysis and the consequent data processing procedures, the final results should be beneficial to the professional surveyor and his associates. It should also prove to be of great value to students studying these subjects for TEC Ordinary and Higher certificates or Diplomas in Land Use, Estate Management and other allied disciplines including CNAA degrees in surveying and management.

By integrating into one book the principles of land measurement, the statistical analysis procedures available and a summary of the data processing services that may be used by a modern office, the reader can appreciate the interrelated aspects of each constituent factor rather than cross-referencing a number of information sources.

The surveying section has been presented in a format which allows a reader with no previous experience of the subject to appreciate the basic principles of land surveying. This basic treatment should be adequate for the reader to demonstrate these principles by undertaking a basic field survey and recognise the applications and limitations of field survey data. The general surveyor must be aware of the map services available and reference to these facilities has been included. The illustrations

Preface

include line sketches to clarify the subject and these should prove useful for examination purposes.

Within the subject of statistics and data processing learning by example is the most appropriate way of attaining a grasp of the subject. Statistical and data processing techniques are best understood by using them, rather than by committing them to memory. In this book, the examples used fall into two categories: simple ones to illustrate a single point or introduce a new method, and more difficult ones comparable with those set in professional examinations.

A particularly important feature of the book is the inclusion of a number of exercises at the end of each chapter to cover the subject matter, and for many of the calculation type questions full answers are provided.

ACKNOWLEDGEMENTS

The authors wish to thank the following:

The Royal Institution of Chartered Surveyors for permission to reproduce questions from the Institution's examination papers.

The Cambridge University Press for permission to reproduce Table I from the *Cambridge Elementary Tables* by Lindley and Miller.

The Technical Press for permission to reproduce an extract from *Tacheometric Tables* by D. T. F. Hunsey in Ch. 9.

Building Research Establishment for the technical information provided from the following Crown copyright publications, by permission of the Controller, HMSO.

BRE Digest 234: *Accuracy in Setting Out*
Current Paper 15/77: *Accuracy Achieved in Setting Out With the Theodolite and Surveyor's level on Building Sites.*

The Ordnance Survey for permission to include information regarding, the national map referencing system of Great Britain, extract from a sample bench mark list and the technical information regarding map accuracy used in Ch. 2.

The kind permission of the following technical sources and publishers is also acknowledged:
Apple Computers (UK) Ltd
C. F. Casella & Co Ltd
Compower (The National Coal Board computing service)
The Estates Gazette Ltd
The Geographical Association
Granada Publishing Ltd
Hall & Watts Ltd
Sokisha Ltd
Wild Heerbrugg (UK) Ltd

Acknowledgements

Carl Zeiss (Zena) Ltd
Oxford University Press

The authors are grateful for the assistance given by a number of colleagues for their contribution, including Mr Tom Johnson for his helpful suggestions on surveying, Mr John Donn for his assistance with the artwork, and our typist Jackie for her patience.

1 LAND SURVEYING

1 LAND SURVEYING PROCEDURES

1.1 INTRODUCTION

The land surveying procedures to be discussed in the following chapters will detail the activities required when identifying the relative positions of both natural and man-made features on the surface of the earth, along with the consequent techniques of recording the field information. The routines for plotting the map or section from these recordings are also discussed. While detailed techniques are available for determining the absolute positions of features on the earth's surface, these have been considered to be outside the requirements of most land surveyors and are discussed fully in other books if further information is required.

1.2 THE CLASSES OF SURVEY WORK

The principal branches of surveying to satisfy the requirements of the land surveyor include:

(a) chain surveying;
(b) traverse surveying;
(c) tacheometric surveying;
(d) plane table surveying;
(e) aerial surveying.

The procedure adopted will be dependent on the application and characteristics of the survey. The following classes of survey work can be identified:

1.2.1 The geodetic survey

This work is concerned with the precise measurement of the position of a feature on the surface of the earth, both in terms of its horizontal

The principles of surveying

location and height above sea level. These procedures are outside the considerations of this book, since astronomical observations and an understanding of the corrections for the earth's curvature are required. It is from this level of accuracy that control points for lower classes of survey work are identified and these are used for a wide range of land survey applications.

1.2.2 The topographical survey

Many of these maps are produced by the Ordnance Survey and indicate the location of both natural and man-made features, with careful consideration given to the representation of ground relief. The method used to identify a feature will depend on the map scale, since this can range from 1/1250 to 1/125 000 for the more popular applications. The use of standard symbols to identify particular features is an important consideration. Larger scales than the above may be used where more accuracy is required regarding the location of detail.

1.2.3 The cadastral survey

This form of survey is undertaken to identify the legal boundaries of specific plots of land and the value of the enclosing area with the ownership and use clearly identified. The level of detail required for this form of map will usually demand a larger scale than the applications above.

1.2.4 The engineering survey

Engineering surveys are usually related to a specific project where information is required at the design stage. The survey will seldom require the absolute position of a feature to be identified, although, a high level of accuracy may be demanded of the work, and this will influence the scale and procedures selected.

Prior to commencing construction work, the features of the project have to be located on the ground and this activity is called setting out. **Setting out** is the reverse procedure of surveying since features are located on the ground from information identified on a map or plan.

1.3 THE PRINCIPLES OF SURVEYING

The basic principles of surveying apply to all classes of work and a detailed understanding of the mathematical principles will be required at the higher levels. These principles are based on the fact that if a triangle is measured on the ground, it can be reproduced to any desired scale providing the distance between two of the points is accurately measured to form a **base line** and one of the following is also measured.

Land surveying procedures

(a) *Two other distances to the third point.* These distances may be measured with a chain or tape for the lower orders of accuracy, although electro-magnetic distance measuring equipment will allow large distances to be accurately measured over terrain where a chain or tape could not be used. This procedure is called trilateration when it includes the measurement of critical angles for checking the work and is used for geodetic surveys. The lower order of accuracy obtained with the chain and/or tape provides the basis of the chain survey used for many construction and development plans. These procedures do not require angular measurements to be undertaken which means that the field work and subsequent plotting procedures are simplified.

(b) *One other length and the included angle with the base line.* Angles are normally observed to a higher order of accuracy than linear measurement observations, therefore a more accurate survey can be produced. These procedures form the basis of the **traverse survey**. Due to the high order of accuracy obtained with the field measurement procedures it is important that co-ordinates rather than angular measurements are used when plotting the survey, to reflect the accuracy of the field work.

(c) *Two angles measured from the base line.* The third point can be located from the base line by angular measurement and these procedures of triangulation formed the basis of **geodetic surveys** until electro-magnetic distance measurement was introduced. By measuring angles rather than linear distances, access is not required onto the intermediate terrain which may be an advantage with many applications, although clear vision of each station must be possible.

1.4 SURVEY ACCURACY AND CONTROL

All survey work must be related to a framework of control points that have usually been fixed to a level of accuracy higher than that demanded by the survey under consideration. The greater the accuracy of the work, the greater will be the cost; therefore, the tolerance demanded will be related to:

(a) the application of the survey;
(b) the scale of the plan or map;
(c) the justification of the cost incurred.

To maintain the desired accuracy it is important that checks are incorporated into the survey at all the stages of the work, including the following:

(a) field measurement; (c) calculations and data reduction;
(b) data recording; (d) plotting.

Survey accuracy and control

In order to undertake the above checks it is often necessary to record observations which do not add to the survey, eg the measurement of angles when undertaking a trilateration exercise. Careful consideration must be given to these requirements when commencing a survey. Table 1.1 outlines the basic classes of error that must be identified when undertaking a survey and the effect that each may have on the completed survey.

The scale to be adopted when plotting the survey may limit the level of detailing that can be identified. Roads cannot be drawn to a scale which is smaller than 1:10000 and the subsequent level of detail demanded in the field is reduced since conventional symbols will be adopted on the map or plan.

Example 1.1

Assuming it is possible to plot to an accuracy of 0.3 mm what distance will this represent at the following scales:

$1:50000 = (0.0003 \times 50000) = 15$ m
$1:10000 = (0.0003 \times 10000) = 3$ m
$1:1250 = (0.0003 \times 1250) = 375$ mm
$1:500 = (0.0003 \times 500) = 150$ mm

Table 1.1 Types of error

Type	Size	Cause	Correction
Blunder	Usually large and random	Personal mistake	Adopt a systematic procedure of work which incorporates checks at each stage
Constant	Proportional to unit of measurement	Incorrect unit of measurement	Correction formulae applied to survey data
Systematic	Irregular in magnitude	Error in field procedure	Check and caution. Adopt procedures which eliminate the effect, eg principle of reversal
Periodic	Positive and negative	Error on scale of equipment	These cancel out with correct procedures which use all the scale range for each observation
Random	May be numerous, should be small and both positive and negative	Error in procedure	Usually cancel each other out when measurement repeated under same conditions

Land surveying procedures

1.5 SURVEY AND PLOTTING PROCEDURES

The following basic procedures apply to all the available survey methods since a systematic routine must be followed, ensuring that each stage of the work is completed before proceeding further, with the related control requirements satisfied. This routine will include the following:

1. Identify the objectives of the survey and the level of accuracy required.
2. Determine the procedure to be adopted and the equipment required to satisfy the objectives identified above. Existing maps, etc. of the area may be used at this stage to identify any difficulties that may occur with the practical fieldwork.
3. Check the equipment that will be required for the work and identify any subsequent compensations that may have to be applied to the data and survey procedures.
4. A reconnaisance of the site is the first stage of the fieldwork to finalise the survey procedure and identify the detail to be plotted. A rough sketch should be completed at this stage to finalise the procedure and identify the controls that will be required.
5. Locate the framework of the survey, ensuring that all the relevant detail can be identified and the respective checks are clearly defined. All the stations should be clearly marked either by wooden pegs or with a more permanent form of monument. Sufficient information should be noted in the fieldbook to relate each station to a permanent object (**witnessing**) at this stage.
6. Field notes must be completed for each section or stage of the work with all the information noted in a clear and concise format. The steps undertaken should provide checks to ensure that all the information has been recorded and is within the accuracy demanded by the objectives.
7. The computation of the results must be neatly set out and undertaken in a consistent manner. This is equally important for manual or computer systems if errors are to be efficiently identified.
8. When plotting the survey the first stage is to locate the framework on the paper. This can be done by either co-ordinates for accurate work, where the used of protractors for angular measurement is eliminated, or by use of geometric shapes for work of a lower accuracy. Any checks included in the fieldwork may then be applied to this framework in order to determine the accuracy of the work. From this framework the detail may be plotted and any symbols or abbreviations included.

A detailed mathematical analysis of these procedures is given in *Fundamentals of Survey Measurement and Analysis* by M. A. R. Cooper published by Crosby Lockwood Staples and the basic application of these procedures is discussed in the following chapters.

1.6 EXERCISE

1. Give one example of each of the following types of error and indicate how the surveyor should deal with it.
   (a) Gross
   (b) Cumulative
   (c) Random
   (d) Instrument
   (e) Permissible

   (RICS)

2 MAP PROJECTIONS AND THE ORDNANCE SURVEY

2.1 SELECTION OF A MAP PROJECTION

When large areas of land are to be mapped on a flat surface, consideration must be given to the accuracy of the final plot in order that:

1. The area is representative of the total area surveyed. This form of representation is called **equal area** or **equivalent**.
2. The shape of the land mass is correct.
3. A constant scale is maintained over the whole area represented. A map which satisfies this requirement is called **orthomorphic** or **conformal**.
4. The bearings between two stated locations are correctly represented.

Since the earth is spherical it cannot be developed into a true flat plane in the same way that a cone or cylinder may be cut and folded flat. This property of the sphere will require the projection used for a specific application to be carefully selected, since no one projection system will satisfy all the above demands. The choice of a particular projection will depend on:

(a) the size and shape of the land mass;
(b) the location of the land mass on the earth's surface and its orientation;
(c) the application of the map, eg navigational, statistical;
(d) the number of plan sheets required to cover the area.

(See *Maps and Diagrams* by F. J. Monkhouse and H. R. Wilkinson, published by Richard Clay, The Chaucer Press Ltd.)

2.2 MAP PROJECTION SYSTEMS

These may be classified according to the following general groups,

2.2.1 Perspective projections

These are formed by conventional geometric projections with the earth's surface drawn as it would be seen from a fixed point. The area under consideration may be projected from either:

(a) the pole, when the centre of the projection will coincide with either the North or South pole;
(b) the equator, when the centre of the projection will coincide with the equator;
(c) at any other desired location, including infinity, to form an oblique projection.

By placing the viewpoint at different locations the following basic projections may be developed.

Orthographic projections. These are generated when the viewpoint is placed at infinity (Fig. 2.1) and will produce a plan which is neither equal area nor conformal. The scale of the projection will be larger near the centre than the outer edge, where it is contracted. The projection has little practical application other than for illustrating books, since it provides the visual effect of a globe but satisfies none of the demands listed in section 2.1.

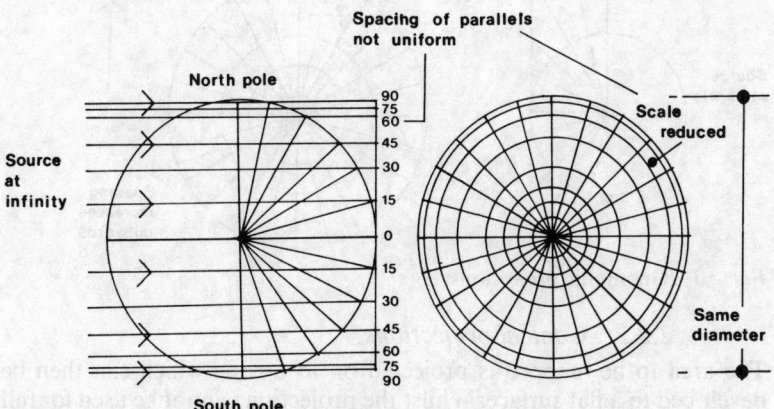

Fig. 2.1 Orthographic projections

Stereographic projections. The projection source or viewpoint is located on the surface of the globe at a point opposite to the area under consideration (Fig. 2.2). This projection produces a map which is larger than the original area and will be **orthomorphic**. While the scale is the same in all directions from a point source, it will increase towards the outer edge. This is a popular projection for polar regions when the viewpoint is placed at the opposite pole and is used for geodetic surveys at a scale of 1:1 000 000 or for meteorological charts.

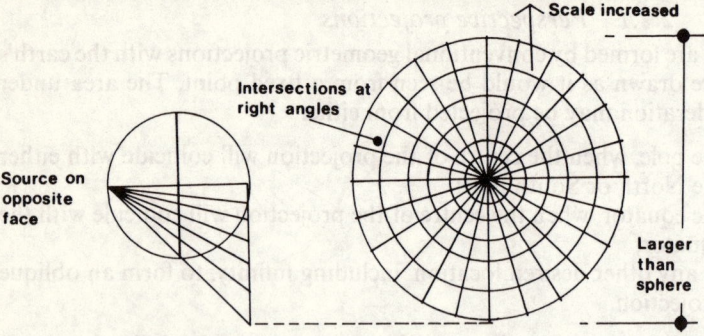

Fig. 2.2 Stereographic projections

Gnomonic projections. The rays are drawn from a viewpoint at the centre of the earth, producing a plan with large distortions in the shape of the land mass. The only advantage of this projection is that the great circles are represented by straight lines and this is popular for sailing charts.

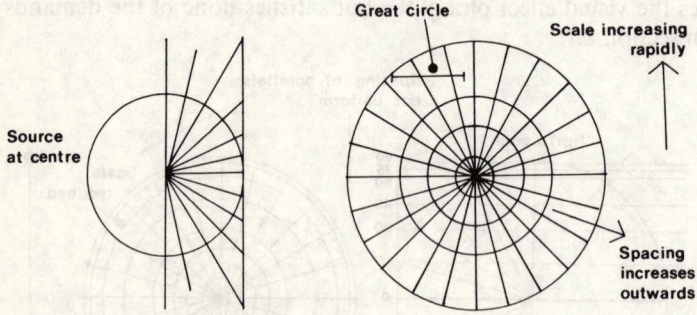

Fig. 2.3 Gnomonic projections

2.2.2 Conical projections

The area to be mapped is projected on to a cone which can then be developed to a flat surface. Whilst the projection cannot be used for full hemispheres or land masses at high altitudes it is useful for topographical and atlas maps since:

(a) the meridians are straight lines converging to the pole;
(b) all latitudes are arcs of concentric circles.

When the cone is placed over the earth with its apex over the north pole, contacts will be made with one parallel of latitude. This circle of contact

Map projection systems

is called **the standard parallel** which passes through the centre of the area to be mapped. The standard parallel (Fig. 2.4) will be of the correct scale and is divided into equal parts by the meridians of longitude, but the scale error will increase with the distance away from the parallel.

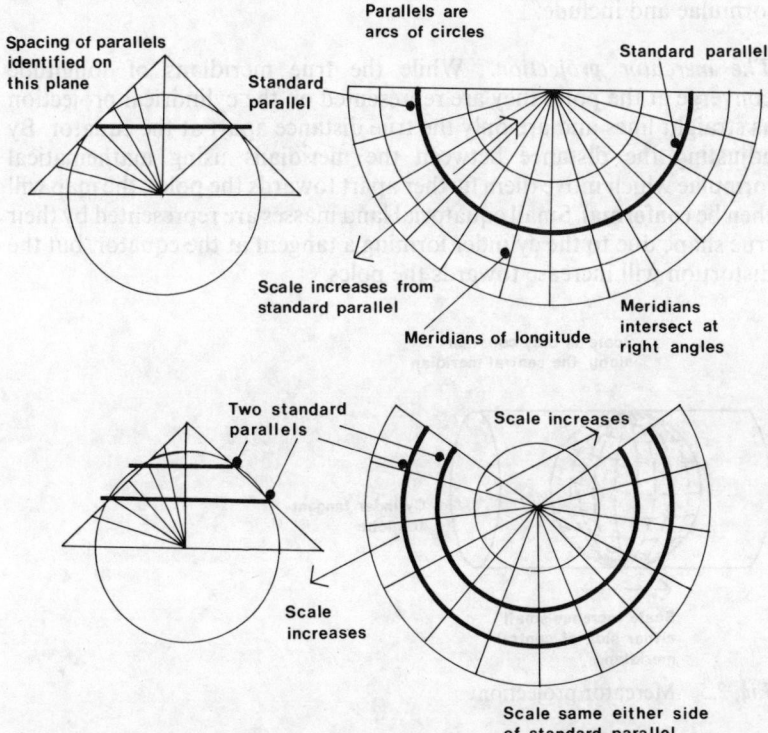

Fig. 2.4 Conical projections

This scale error may be modified by passing the cone through **two standard parallels** each represented at the true location. The resulting scale error will increase north and south of the standard parallels but diminish between them. This will produce a useful map with only small errors of scale and shapes if the standard parallels are carefully selected. Unfortunately it will be neither equal area, nor orthomorphic.

Bonne's projection is a modified conical projection using one standard parallel and is designed to represent the areas correctly with the correct perpendicular distance between parallels.

2.2.3 Cylindrical projections

The features are projected on to a cylinder which can then be developed to produce a flat map. The basic projection places the cylinder at a tangent to the equator and this forms the standard parallel. The meridians of longitude are represented as straight lines spaced equally along the equator, although the parallels of latitude are considerably exagerated away from the equator. Since the basic projection is of little use, most cylindrical projections have been adjusted using mathematical formulae and include:

The mercator projection. While the true meridians of longitude converge at the poles they are represented on the cylindrical projection as straight lines and are only the true distance apart at the equator. By adjusting the distance between the meridians using mathematical formulae which move them further apart towards the poles, the map will then be **conformal**. Small equatorial land masses are represented by their true shape due to the cylinder forming a tangent at the equator, but the distortion will increase towards the poles.

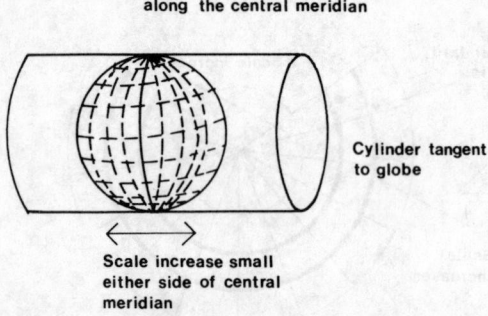

Fig. 2.5 Mercator projection

When the equator is used as the tangent for the projection, any straight line joining two locations will be of constant compass bearing and this is called a **rhumb line**. **Great circles** identify the shortest distance between two locations and these will be curved. Except for navigational purposes, where the above characteristics are beneficial, the projection has few attributes other than for equatorial locations, since the land mass is distorted.

The transverse mercator projection. This projection places the cylinder at a tangent to the globe, along any chosen meridian. This will enable a narrow area of land extending a few degrees either side of the meridian to be a true conformal projection, with only a small change in scale over a large distance. The meridians of longitude and parallels of latitude are each orthogonal curves intersecting at right angles, with the central meridian a straight line (Fig. 2.5).

If the cylinder is smaller than the earth it intersects the earth twice instead of once and the scale will then be constant along both these cut lines. This will provide a useful projection for topographical maps at a large scale.

The **transverse mercator projection** is now used for all new maps of the United Kingdom with the central meridian, coinciding with longitude 2° West and running north to south, since this is the principal direction of the land mass. Although the scale is the same in all directions at a location, it will increase by approximately the square of the distance from the central meridian to a maximum error of 1/1250 at the eastern (Norfolk) and western (Cornwall) extremities. The transverse mercator projection minimises this error by reducing the scale by 1:2500 at the eastern and western extremities; this enables the correct scale to be obtained along the parallels 180 km either side of longitude 2° West.

The scale factor. The local scale factor at a location is expressed by the ratio of:

$$\frac{\text{grid distance}}{\text{ground distance}}$$

This factor is assumed to be constant for areas of land up to 10 km radius, which includes the majority of survey applications. Tables of local scale factors are available for each 10 km of the National Grid (section 2.3). Due to the reduction of the scale on the meridian a scale factor of 1 is found 220 km and 580 km east of the National Grid origin.

Direction correction. A line of sight between two points will be represented by a curved line on the projection. This curvature will be a maximum of 7 seconds when the line runs north–south and zero when the line runs east–west. The correction will be added if the line is west of the meridian and subtracted if it is east. This correction is only required for special applications outside the range of procedures discussed here.

2.3 THE NATIONAL GRID

2.3.1 Objectives of the National Grid

Following the final Report of the Departmental Committee on the Ordnance Survey (the Davidson Committee), a grid referencing system was devised to satisfy the following objectives:

1. Provide a unique referencing system which will enable a feature to be accurately located on a map.
2. The grid reference of a feature on one map will be the same for all maps irrespective of the scale.

The grid is a series of lines drawn parallel and at right angles to the central meridian of the transverse mercator projection and covers England, Scotland and Wales.

2.3.2 The method of grid referencing

The whole of the land mass is included in a rectangle measuring 1300 km × 700 km so that any location can be referenced to identify its position. These co-ordinates do not relate to any rectangular co-ordinates derived from the transverse mercator projection because they are not derived from the same origin and the grid does not fit the meridians of longitude. The relationship between the referencing grid and the meridians of longitude is called **convergence** and special tables are available for this correction, but they will not be required for the survey procedures discussed here.

The origin of the transverse mercator projection is 2° West and 49° North while the false origin of the National Grid is 7° 33′ West 49° 46′ North, which is approximately west of the Scilly Isles. This point has been chosen so that any point under consideration will be **east** and **north** of the origin. Had the origin of the transverse mercator projection been used then some locations would have been west of this point. The independence of the National Grid as a referencing system means that any references used do not relate to actual distance but to a location **east** and **north** of the **origin**, as illustrated in Fig. 2.6.

The grid consists of six 500 km squares designated by the letters S, N, H, J, O, T which are then sub-divided into 25 further squares of length 100 km each given a letter of the alphabet excluding I. This system means that no two squares can have the same combination of letters and a location can be identified to within 100 km, as indicated below:

London TQ
Bristol ST
Manchester SJ

Within each 100 km square a location is identified by its distance from the **south–west** corner of that square. The reference co-ordinates are stated by the distance eastwards and then the distance north from the reference corner, such that NQ 26 will be a location in the 100 km square designated NQ 20 m east and 60 m north of the south–west corner of square NQ. (See Fig. 2.6.) A point location within a 10 km grid square can be further identified by estimating the tenths of the respective 10 km square as illustrated in Fig. 2.6.

When quoting a grid reference the distance east is quoted first followed by the distance north, each being recorded to the same level of accuracy. This means that all grid references will contain two letters followed by an even number of digits, as follows:

The National Grid

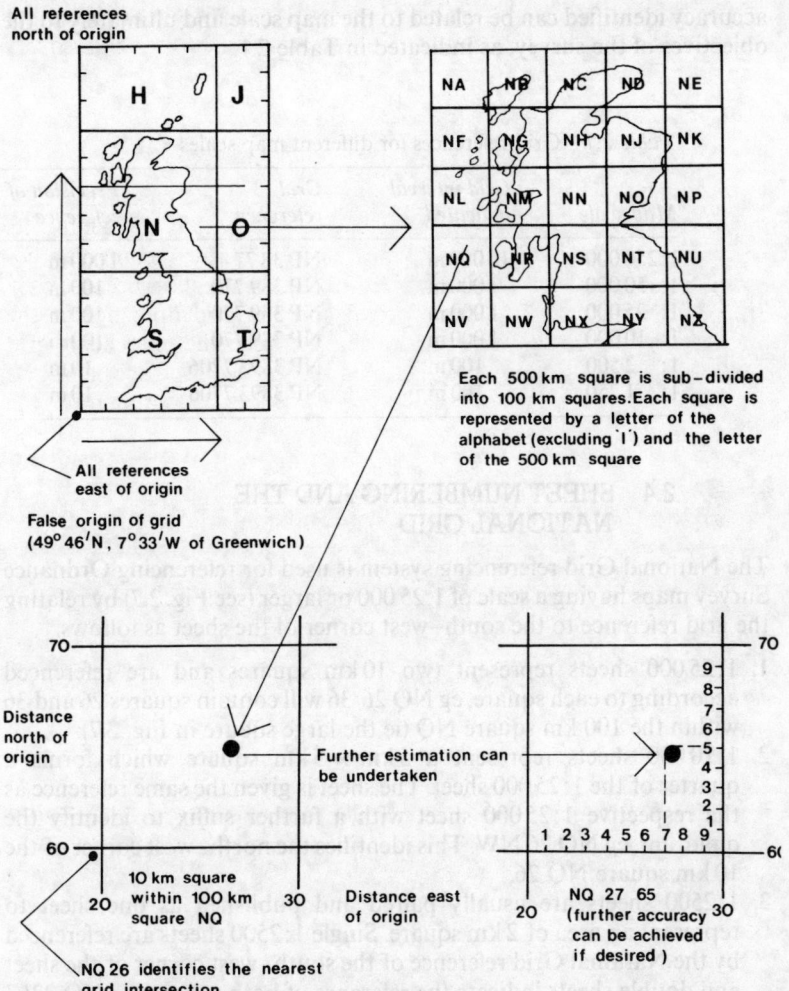

Fig. 2.6 The National Grid referencing system

NQ 27 65

This reference identifies the following:

N identifies the 500 km square;
Q identifies the 10 km square within the 500 km square.
27 identifies the distance east measured in kilometres from the south-west corner of the 10 km square.
65 identifies the distance north in kilometres from the same location.

The system ensures that each reference is unique and the level of accuracy identified can be related to the map scale and ultimately to the objectives of the survey, as indicated in Table 2.1.

Table 2.1 Grid references for different map scales

Map scale	Grid interval indicated	Grid reference	Precision of reference
1:250 000	10 000 m	NP 38 77	1 000 m
1: 50 000	1 000 m	NP 389 770	100 m
1: 25 000	1 000 m	NP 389 770	100 m
1: 10 000	1 000 m	NP 389 770	100 m
1: 2 500	100 m	NP 3893 7706	10 m
1: 1 250	100 m	NP 3893 7706	10 m

2.4 SHEET NUMBERING AND THE NATIONAL GRID

The National Grid referencing system is used for referencing Ordnance Survey maps having a scale of 1:25 000 or larger (see Fig. 2.7) by relating the grid reference to the south–west corner of the sheet as follows:

1. 1:25 000 sheets represent two 10 km squares and are referenced according to each square, eg NQ 26/36 will contain squares 26 and 36 within the 100 km square NQ (ie the large square in Fig. 2.7).
2. 1:10 000 sheets represent a 5 km × 5 km square which forms a quarter of the 1:25 000 sheet. The sheet is given the same reference as the respective 1:25 000 sheet with a further suffix to identify the quadrant, eg NQ 26 NW. This identifies the north–west corner of the 10 km square NQ 26.
3. 1:2500 sheets are usually paired and published as one sheet to represent an area of 2 km square. Single 1:2500 sheets are referenced by the National Grid reference of the south–west corner of the sheet and double sheets indicate the reference of both squares eg NQ 2267 and NQ 2367.
4. 1:1250 sheets represent an area of 500 m square with four 40 cm × 40 cm sheets representing a 1 km square. The sheets are referenced in the same format as the 1:2500 sheets with the inclusion of a quadrant reference, eg NQ 2267 SE.

2.5 MAPS PRODUCED BY THE ORDNANCE SURVEY

The Ordnance Survey provide a wide range of maps commencing with the 1:1250 and 1:2500 scales which clearly identify the physical detail and parliamentary boundaries in black and white, to specialised small-

Maps produced by the Ordnance Survey

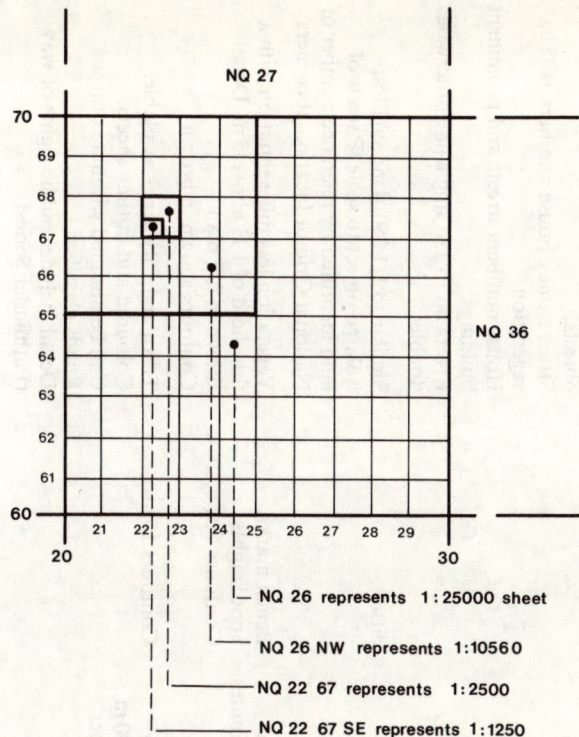

Fig. 2.7 Sheet numbering and the National Grid

scale maps utilising a high level of colour, and hatching to identify relief, symbols to identify physical detail, etc. A summary of the characteristics of the maps most popular with a surveyor are given in Table 2.2.

Many of the maps outlined in Table 2.2 have been inherited from older surveys and have been either re-surveyed or subject to revision where errors have been identified. Advanced Revision Information (ARI) is available where revision of a map has taken place but the new edition has not been published. Provision is also made for customers to call at local Ordnance Survey offices to examine the Master Survey Drawings (MSD) of the area. These are the working documents of the locally based surveyor and include the current field notes. Other services include:

1. Information regarding triangulation stations and minor control points in urban areas.
2. Levelling information in the form of bench mark lists.
3. Aerial photographs at various scales in the form of prints, enlargements or negatives.

Table 2.2 The characteristics of Ordnance Survey maps

Scale	Format	Reference	Grid	Levels	Features
1/1 250	0.5 km sq quadrant of 1:2 500 400 mm square covers 500 m × 500 m	SJ 8975 NE	Grid 100 m margin 10 m	Bench marks Spot heights	Cover urban areas. Mainly man made, permanent objects. Street names, house numbers, 0.25 ha vegetation. House numbers streets and prominent buildings. Rivers up to 1 m single line otherwise double.
1/2 500	2 km × 1 km Double km sheets	SJ 8975–9075	ditto	ditto	Similar to 1/1 250, 4 m² smallest isolated object to scale. Parcels of land identified by reference number of National Grid or location at corners
1/10 000	cover country 5 km sq Quadrant of 1:25 000	SJ 87 NE	Grid 1 km margin minutes	Bench marks Spot heights Contours 5 m or 10 m	Vegetation less differentiation with a threshold of 1 H above 5 m. Detail shown in full plan. Contours shown in brown
1/25 000	20 km × 10 km 10 km × 10 km 10 km × 14 km (Coastal)	Torbay SX 86/96	Grid 1 km margin 100 m and minutes	Contours 10 m	Town and City series available. Coloured and outline sheets. Can be outlined without contours. Roads classified. Detail includes public rights of way (Pathfinder Series)

1/50 000	40 km sq	204 cover the country (numbered)	Grid 1 km margin minutes	Contours 15 m or 16 m and spot heights to nearest 1 m	Public buildings black, others orange. Civil boundaries correct for London only. Roads coloured
1/100 000	varies	county name			Administrative areas and petty sessions
1/250 000	190 km × 140 km	9 sheets numbered 1 to 9	Grid 10 km margin 10 min 30 min. crosses	Spot heights Contours	Roads, railways, relief layered, hills can be outlined, Scotland administrative, gazetteer
1/625 000	1 sheet with H and S printed back to back when folded or separate.		Grid 10 km margin 5 min.	Hill shading	Route planning official record of trunk roads with submarine layering. Topographical base with outline, black, no relief, contours

19

Map projections and the ordnance survey

4. An enlargement and reduction service on film or paper.
5. Digital mapping of new editions of the 1:1250 and 1:2500 maps. This is available on computer tape which may be used in conjunction with the Ordnance Survey Fortran Customer Plot Program. This program will plot data at a range of scales, plot selected information and plot a rectangular-shaped area ignoring map sheet edges.
6. Other facilities include a mounting and finishing service for specialised work.

2.6 PARCEL NUMBERS

The 1:2500 map is divided into parcels of land for the surveyor to identify the area of land plots, etc. Each parcel is given a reference number which is the four figure grid reference of the centre with the area in hectares and acres. A parcel which includes the corner of a map is given a number from 0001 to 0006 so that any number will not be repeated on other maps which surround the parcel. Where adjacent features are included in a parcel this is known as a brace and is denoted by a brace symbol, ∫, or when a parcel is defined by a double feature with a central boundary or where the boundary is not more clearly defined, then other symbols (see Fig. 2.8) are used.

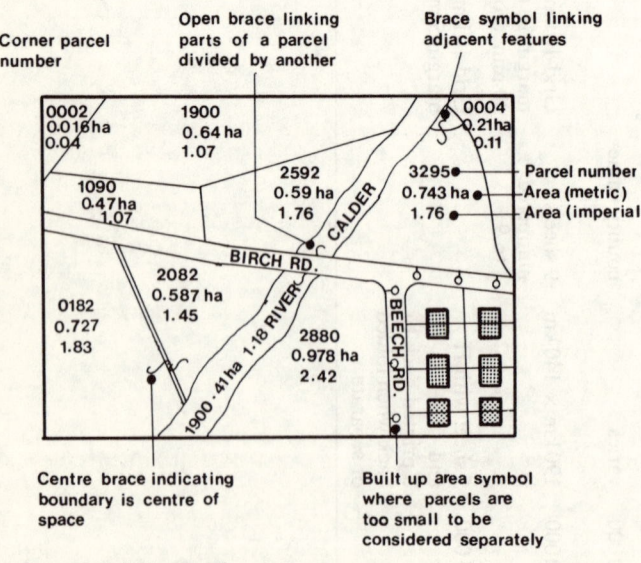

Crown copyright reserved

Fig. 2.8 Parcel numbers (adapted; courtesy Ordnance Survey)

2.7 THE ACCURACY OF ORDNANCE SURVEY MAPS

2.7.1 Introduction

The accuracy of Ordnance Survey maps is limited by the number of old maps that have been inherited, the cost of detailed survey procedures and the requirements of modern map production techniques. These demands mean that whilst some maps are produced by a re-survey of the required area (1:1250 covering towns, etc. with a population over 20 000); others are revised and overhauled. Where revision work has been undertaken Advance Revision Information sheets are provided to update existing maps.

Since accuracy is relative to the application of a particular map sheet, no rigid definition or limit can be given although it will be recognised by the definition of:

(a) topographical detail;
(b) description of detail;
(c) horizontal location of detail;
(d) altitude;
(e) National Grid referencing.

This level and range of parameters indicates that a map user must carefully select a map relative to the application being considered, since one map cannot satisfy all the above demands, even though controls are applied in the following manner.

2.7.2 The level of detail

The scale of a map will limit the level of detail that can be clearly located without the use of symbols, as follows:

Scale	Smallest clearly defined detail
1:1250	0.25 H
1:2500	0.4 H
1:10 000	I H

2.7.3 Descriptions

Where place names are mentioned in Acts of Parliament they are included; where names have derived from local custom or common usage then these will be used, after local enquiries have been made. Standard lists of descriptive terms are also produced for consultation, when the specific definition of a particular term is required. Any changes in public boundaries or the introduction of new features, etc. will not be identified until a revision is considered and these may require verification.

2.7.4 Accuracy of map scales

Regular tests are given to 3 per cent of the maps produced by the Ordnance Survey to identify the priority of future revision areas. The checks are carried out to ensure that individual details are represented to the same level of accuracy throughout a map series along with the following test programme:

1:1250 Re-survey and continuous planimetric revision;
1:250 Re-survey and continuous planimetric revision;
1:10 000 Re-survey and continuous planimetric and contour revision.

2.7.5 Accuracy of triangulation

Inaccuracies in the control framework are assessed and measured by regular observations using electro-magnetic distance measuring equipment (section 5.4). These observations have indicated that the primary triangulation may be 15 parts per million too large, while astronomical observations with modern equipment indicate an error of up to 5 seconds. This practice of combining different survey techniques is often used to identify errors in the national survey.

2.7.6 Accuracy of contours

The accuracy of levelling will depend on the procedures used, the datum selected and the scale adopted. The maximum error of a contour should not exceed one quarter the contour interval.

Since spot heights represent staff positions, errors will occur due to the location of the plan position. These are often checked by relating the results of air photographs to the results produced by a microptic alidade and staff. The Ordnance Survey publish the trigonometrical heights of triangulation points in the National Grid Triangulation List. Even though some of these heights are determined by theodolite and related to spirit levelling procedures, the level of accuracy is difficult to determine.

2.7.7 High and low water marks

Where a map identifies the limits of the foreshore it is identified in England and Wales by the mean high and low water location of an average tide; in Scotland an average spring tide is used.

The survey method used is determined by the physical nature of the foreshore and includes direct measurement where conditions are favourable or air photographs using infra-red film where conditions are difficult. The identification of low water is often difficult to establish even with the above techniques and cannot be as accurate as the clearly defined high tide location. The level of both the high and low water will depend on the coastal location and tide characteristics therefore this line with regard to the Ordnance Datum (section 8.1) cannot be considered as a contour line.

2.7.8 Accuracy of National Grid co-ordinates

Randomly selected points of detail are identified and the co-ordinates computed from field surveys or aerial triangulation. By modifying the survey procedures and selecting up to 500 or more test locations, the source of any error can be evaluated.

The checks are undertaken by an order of survey accuracy higher than the original survey and often to a larger scale; for example, the co-ordination of check points when overhauling the 1:2500 maps was undertaken by aerial triangulation at a scale of 1:7500 with ground control stations located by instrument methods. The errors identified by the above procedures are then established where further investigation and possible map revision is required.

(Further information regarding the range of mapping services and accuracy of Ordnance Survey maps may be found in *Ordnance Survey Maps, a Descriptive Manual*, by J. B. Harley, Oxford University Press for H.M.S.O.)

2.8 EXERCISES

1. Describe with the aid of sketches how the altitudes of land are shown on 1/2500 and 1/10 000 scale maps of the Ordnance Survey. These altitudes are referred to a certain datum. What is this datum and which is the site where observations are made for the datum?
 (RICS)
2. The Ordnance Survey provides coverage of the British Isles by a series of maps at different scales with a unified National Grid Reference System.
 Give an account of the mapping describing the detail included at each scale. (RICS)
3. If asked by a client, how would you describe the accuracy of Ordnance Survey maps? (RICS)
4. Describe the National Grid in the context of the Ordnance Survey and illustrate the reference system and state the principal uses in surveying practice of the various scales of maps available. (RICS)

3 THE CHAIN SURVEY

3.1 INTRODUCTION

This form of survey will enable features to be plotted on a plan from the direct measurement of horizontal distances and may be used to produce a plan of a small area. The procedures have the advantage that they demand no calculation for work of a low accuracy, require a minimum of equipment and the procedures used are simple to carry out, both in the field and the subsequent plotting. Since it is necessary to measure horizontal distances in the field, it is important that the area between consecutive stations is accessible and capable of practical linear measurement. This may restrict the survey procedures for certain applications.

3.2 SURVEY PRINCIPLES

The area to be surveyed must be capable of being divided into a series of triangles to form a control framework that can be plotted to a desired scale. This framework is most accurate if it can be developed from the longest line passing through the survey. This is termed the **base line**.

The topographical or man-made features to be identified are then related to this control framework by further measurement and these should be as short as possible. Care must be taken, therefore, to locate the chain lines close to the features to be identified for maximum accuracy (see Fig. 3.2). Other factors to be considered are:

(a) the base line on which the framework is structured should be as long as possible for maximum accuracy (DB in Fig. 3.2);
(b) all the main triangles should be related to the base line;

(c) there should be as few triangles as possible consistent with covering the area and locating the detail;
(d) triangles should be well conditioned, having angles between 30° and 120° to give clean intersections;
(e) respective stations should be clearly visible;
(f) survey lines should promote ease of ranging and measurement, avoiding obstacles and sloping ground;
(g) all triangles should be checked or proven;
(h) if possible avoid survey lines without offsets unless they are check lines (Fig. 3.1).

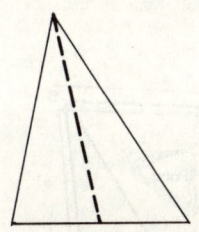 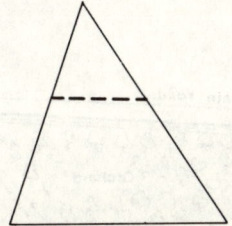

Fig. 3.1 Check lines

The procedures outlined 'work from the whole to the part', and if required smaller triangles or detail lines can be constructed within the framework as demanded by the survey (Fig. 3.2).

3.3 SURVEY EQUIPMENT

The principal equipment for linear measurement is the steel or galvanised iron chain which is suitable for low accuracy work and has the advantage of being very robust. For more accurate work a range of tapes and steel bands are available. The equipment is specified in BS 4484: PART I: 1969 *Specification for measuring instruments for construction work* and is outlined in Table 3.1.

Other equipment required during the survey will include:

Ranging rods. These are circular poles made of timber, aluminium or steel 2 m, 2.5 m, or 3 m long, painted, having distinctive red and white horizontal bands. The lower end is tipped with a pointed steel shoe to allow it to be driven into the ground. This rod may also be used to measure short horizontal distances using the shoe to locate the tape or the horizontal bands on the staff for lower orders of accuracy where a low order of accuracy is satisfactory.

The chain survey

Arrows. These are steel skewers about 400 mm long with a red ribbon or other distinctive feature located at the top. They are used to locate the end of a chain length (see section 3.5).

Pegs. Timber pegs are used to locate the stations. Metals studs may be used on hard road surfaces, etc.

Magnetic compass. This is used to obtain the bearing of one of the survey lines in relation to magnetic north, so as to orientate the survey. The line selected must be free from magnetic attraction (section 4.2) and is usually the base line.

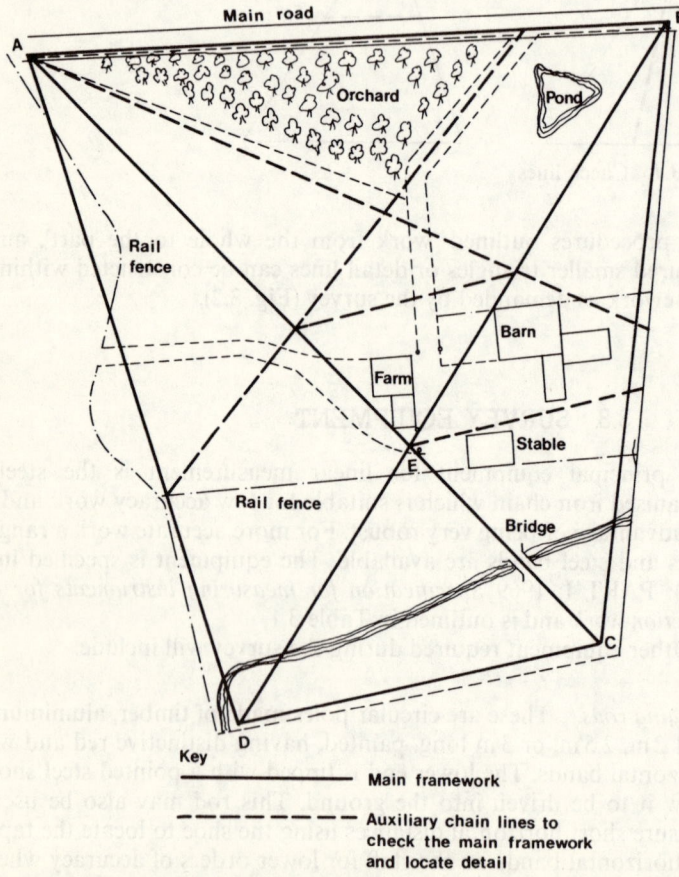

Fig. 3.2 Chain survey framework (RICS)

Table 3.1 Equipment for linear measure
(REF. BS 4484: Part 1: 1969 for measuring instruments for construction work)

Instrument	Material	Length	Graduations	Remarks	Accuracy
1. Chains	Tempered steel wire, consisting of straight lines connected by three oval rings	20 m	Links at 200 mm c/c with yellow tally markers at 1 m and red figured tallies at 5 m c/cs	Robust, easy to read. May be repaired in the field. Care required when folding.	Low accuracy. May be shortened due to bent links, etc.
2. Steel band	Steel strip in a frame complete with winder. Invar bands for improved accuracy	10 m 20 m 30 m 50 m	On upper face only. 1 m grads in colour. Major grads with figs. at 100 mm c/c. Arrows indicate 50 mm intervals.	More accurate than chain. Difficult to repair in the field	Calibration temp and tension should be indicated within first 300 mm. High accuracy
3. Tapes	Steel/vinyl	10 m 20 m 30 m	Major grads and figures at 100 mm. Intervals grads at 50 mm and 10 mm indicated by arrows	Grads easily visible, very stable	Operating tension indicated. Improved accuracy over other tapes
	Fibreglass	As above	100 mm grads figured, 50 mm and 10 mm Intervals indicated	More robust than linen. Easy to clean. Grads easily visible	Stable in length. Low accuracy tension should be indicated
	Linen	As above	As above	Less stable than Fibreglass. Flexible. Easily discoloured	Low accuracy

The chain survey

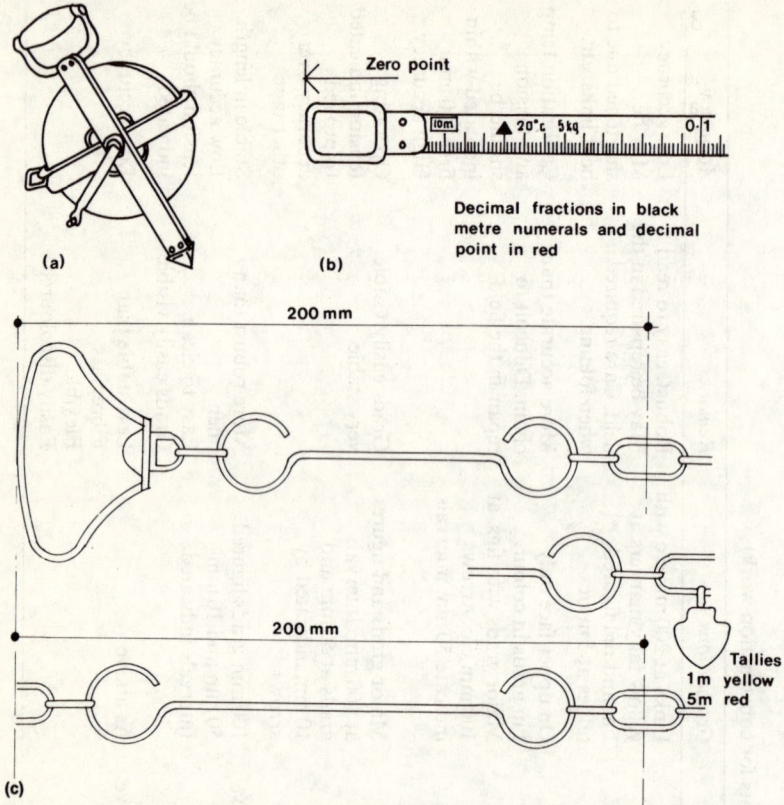

Fig. 3.3 Survey equipment (a) steel band (b) steel tape (c) the land chain

3.4 SURVEY PROCEDURE

1. Reconnoitre the site to note all relevant details, features and characteristics, etc, and decide on the survey method.
2. Identify the location of each station, ensuring that the location can be occupied, all related stations are visible and the triangles formed satisfy the requirements outlined in section 3.2.
3. Sketch an outline of the survey in the first page of the fieldbook, indicating the survey framework and lettering each station. Sketch the features of the framework, taking care to exaggerate any, where this will assist the subsequent plotting.
4. Fix the stations by means of wooden pegs driven into the ground and **witness** them. This requires a sketch of the features adjacent to the station with at least **three** distances from permanent features noted.
5. Chain each survey line and locate the required detail.
6. Booking of detail.

3.5 CHAINING A LINE

This is the term given to the measurement of the distances between the stations and is best undertaken by two people, one called the **leader** pulling the chain along and the other called the **follower** coming up behind. These people undertake the following procedures:

1. The leader holds both handles of the chain in one hand, throws out the chain, opening the chain to its full extent on the ground.
2. While the follower remains at Station A, holding the back of the chain handle against the station, the leader takes a number of arrows, a ranging pole and the other end of the chain towards Station B.
3. The follower lines in the leader along line A–B (Fig. 3.4) by sighting along the line A–B identified by the ranging poles at each station. This procedure is used when a line greater than the chain length is to be measured and is called **ranging a line**.
4. The leader places the ranging pole approximately one link short of the chain length, pulls the chain straight, inserting an arrow outside the chain handle at A1.
5. The leader than drags his end of the chain to A2, taking the remaining arrows with him.
6. The follower brings his end of the chain against the arrow at Station A, which is replaced by a ranging pole. The follower saves the arrow removed since this will serve as a check of the number of chain lengths measured over the total distance.
7. The above procedure is then repeated until station B is reached. The chain is pulled past the ranging pole indicating the station and the short distance to be added to the full chain lengths is noted.

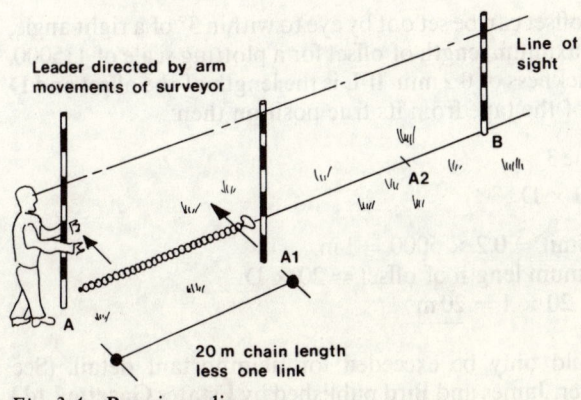

Fig. 3.4 Ranging a line

The chain survey

3.6 THE LOCATION OF DETAIL

The method of locating detail from the chain line will depend on the characteristics of the feature, the accuracy required and its distance from the chain line, but will normally consist of one of the following procedures:

Offsets. These are measurements taken at right angles from the chain line to the feature under consideration using a linen tape for normal work (Fig. 3.5). Where the distance to more than one feature is noted on the same offset these are called **running offsets**.

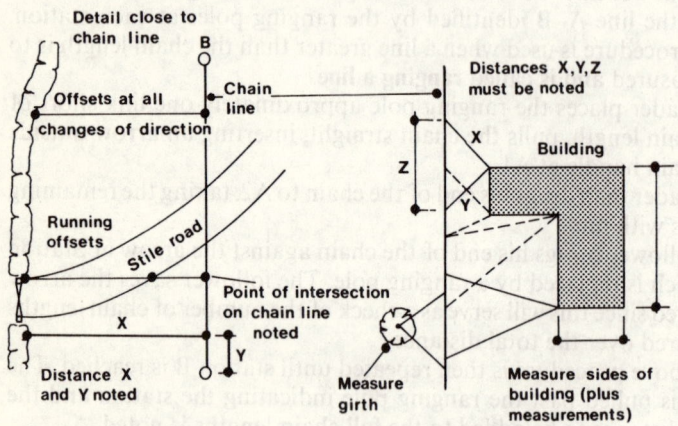

Fig. 3.5   Offset detail   Fig. 3.6   Tie detail

Example 3.1

Assuming that an offset can be set out by eye to within 3° of a right angle, what will be the maximum length of offset for a plotting scale of 1:5000, assuming a line thickness of 0.2 mm. If L is the length of the offset and D the displacement of the tape from its true position then:

$$D/L = \sin 3°$$
$$L \simeq 20 \times D$$

Plotting limit = $0.2 \times 5000 = 1$ m.
The maximum length of offset $\simeq 20 \times D$
$20 \times D = 20 \times 1 = \underline{20\,\text{m}}$

This limit should only be exceeded for unimportant detail. (See *Surveying* by Garner, James and Bird published by Estates Gazette Ltd.)

Setting out right angles

Ties. These consist of two measurements from the chain line to the feature to form a well-conditioned triangle where the lengths of all sides are known (Fig. 3.6). This procedure is used for points of detail or where an offset will be too long for accurate location.

Plus measurements. These are taken in addition to offsets around the perimeter walls of buildings, etc so that the location can be accurately plotted and the offset measurements verified.

3.7 SETTING OUT RIGHT ANGLES

The procedure to be adopted must be selected with regard to:

(a) the purpose of the survey;
(b) the scale used when plotting;
(c) the location of the detail with regard to the chain line.

Example 3.1 illustrates the error that may be expected when manually estimating a right angle, therefore one of the following procedures may be adopted to improve the accuracy of the offset:

3.7.1 Manual procedures

There are a number of procedures available using normal survey tapes, etc, depending on the location of the detail with regard to the chain line:

Right angles from a fixed point on the survey line. Where a right angle is required at C, being a point on the chain line, a 3:4:5 triangle may be formed (Fig. 3.7).

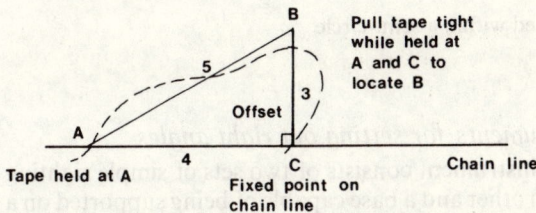

Fig. 3.7 An offset formed using a 3:4:5 triangle

AC is measured along the chain line from point C and the tape extended in the direction of B until the 7 m mark is identified on the tape. Place the 12 m mark on the tape at point A on the chain line and pull the tape tight from the 7 m mark. This will form a right-angled triangle with the line BC at right angles to the fixed line CA.

The chain survey

A fixed location within one chain length of the chain line. Where it is desired to locate an offset from a fixed point A, swing two arcs of equal length to intersect the chain line at CD. The mid point B of the line CD on the chain line (Fig. 3.8) gives the location of the desired offset AB.

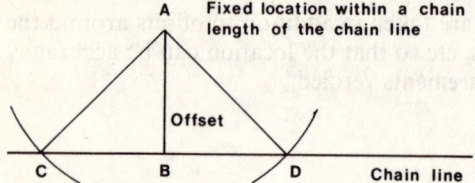

Fig. 3.8 An offset formed by equal arcs

A fixed location located more than a chain length from the chain line. From the fixed point A, range in line AC so that C is on the chain line but distant from the desired offset location and AC is less than two chain lengths long. Bisect AC to give point D and describe a circle of radius AD, centre D, to intersect the chain line at B (Fig. 3.9). AB is then the desired offset.

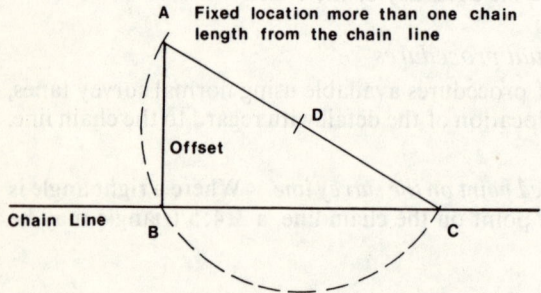

Fig. 3.9 An offset formed within a semi-circle

3.7.2 Instruments for setting out right angles

The cross staff. The instrument consists of two sets of simple sighting vanes set at 90° to each other and a base capable of being supported on a ranging pole at the eye level of the operator (Fig. 3.10).

The optical square. The construction of the instrument is based on the principle that when a ray of light is reflected from a pair of mirrors facing each other, the light ray is redirected through an angle twice that subtended by the mirrors. The arrangement of mirrors indicated in Fig. 3.11 will allow the observer to see a ranging pole placed on the chain line

The fieldbook

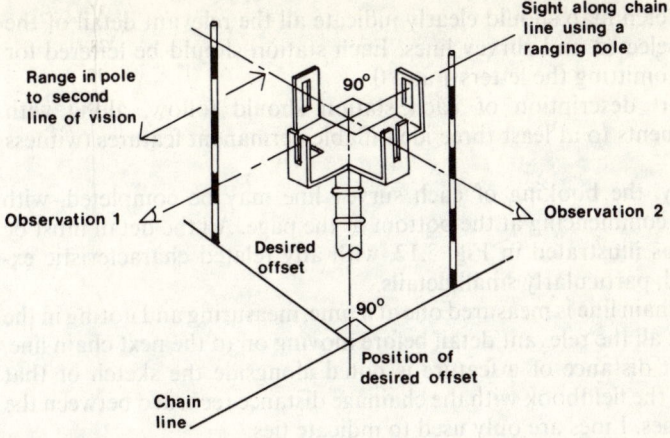

Fig. 3.10 Application of the cross staff

through an aperture in the instrument along with an image of a further ranging pole at 90° to the chain line observed in a mirror placed below the aperture. Modern instruments use a prism instead of a set of mirrors since this is more robust.

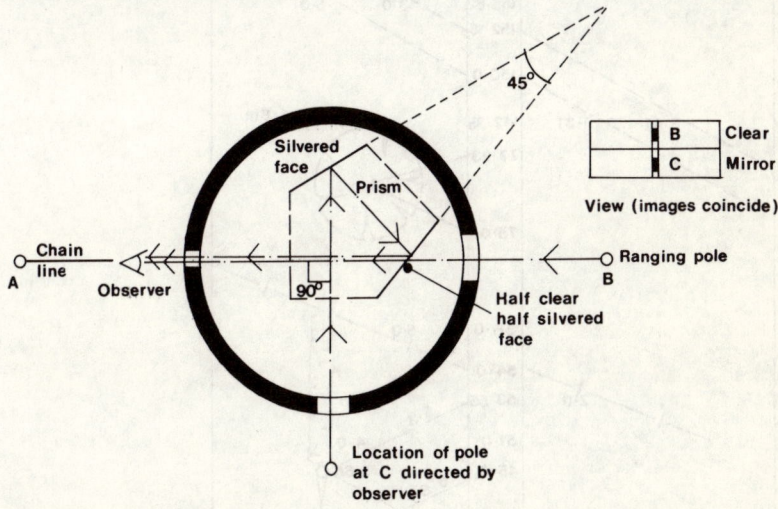

Fig. 3.11 The optical square

3.8 THE FIELDBOOK

The first page of a fieldbook should contain the name or title and location of the survey, the date, names of surveyors and a clear sketch map of the survey area.

The chain survey

The sketch map should clearly indicate all the relevant detail of the stations selected and survey lines. Each station should be lettered for reference omitting the letters 1 and 0.

A short description of each station should follow, along with measurements to at least three identifiable permanent features (witness stations).

Finally, the booking of each survey line may be completed, with each line commencing at the bottom of the page. All the detail must be inserted as illustrated in Fig. 3.12, with any related characteristic exaggerated, particularly small details.

Each chain line is measured one at a time, measuring and noting in the fieldbook all the relevant detail before moving on to the next chain line. The offset distance of a feature is noted alongside the sketch of that feature in the fieldbook with the chainage distance recorded between the double lines. Lines are only used to indicate ties.

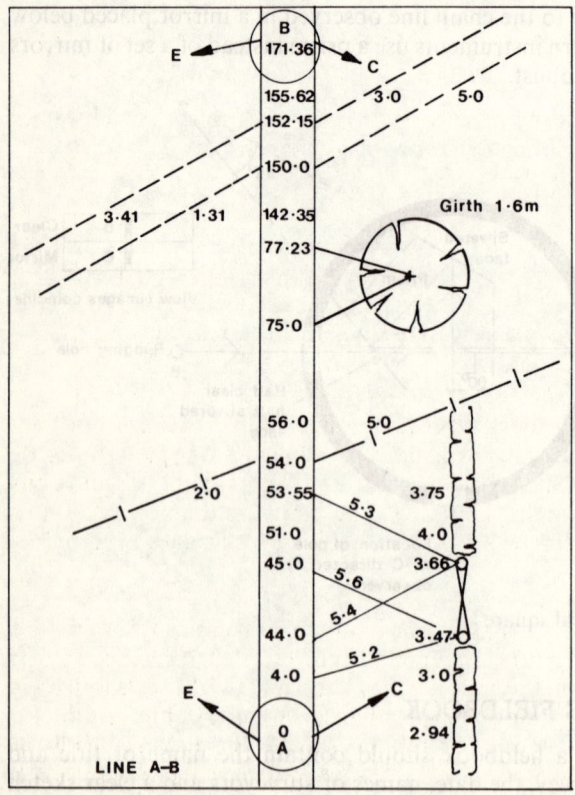

Fig. 3.12 The fieldbook

3.9 CORRECTION FOR SLOPING GROUND

3.9.1 Introduction

The required distance between two known points in surveying is the horizontal or plan distance, while the measured distance in the field is taken as being a line between the stations on the ground surface. Where the ground slopes the measured distance will exceed the required horizontal distance as indicated in Fig. 3.13.

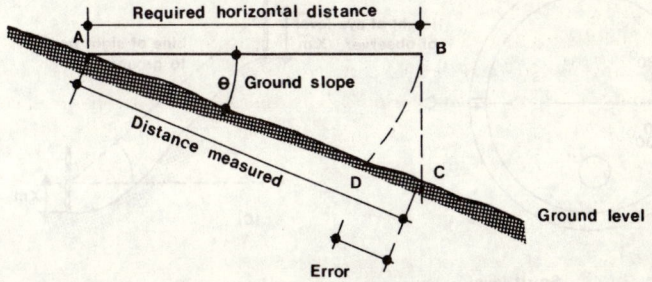

Fig. 3.13 The relationship between the horizontal and sloping distance

Figure 3.13 indicates that the correction DC must be **subtracted** from length AC to obtain the horizontal distance AB.

$$\frac{\text{Required horizontal}}{\text{distance}} = \frac{\text{Measured}}{\text{length}} \times \cos \theta.$$

Note: The correction is always *subtracted* from the measured length.

Up to a slope of 3° and a gradient of 1 in 20 any correction can be ignored due to the low order of accuracy of a chain survey.

3.9.2 Correction procedures

Stepping. This procedure requires no calculation and is often the best procedure where a series of irregular slopes under 5 m in plan length are encountered. The chain is held horizontal in short lengths to reduce any error caused by the chain sagging. The point on the ground is then located by means of a drop arrow and the procedure repeated as outlined in Fig. 3.14(a).

Watkins clinometer. This instrument is used for measuring the angle of ground slope by means of a counter-weighted scale freely suspended so that the 0 division on the scale is always horizontal. The scale is divided from 0° in both **elevation** and **depression**, as outlined in Fig. 3.14(b).

To measure the slope of a ground line AB the surveyor with the instrument stands at Station A (Fig. 3.14c) and an assistant stands at B with a ranging pole having a clearly marked point at the eye level from A;

The chain survey

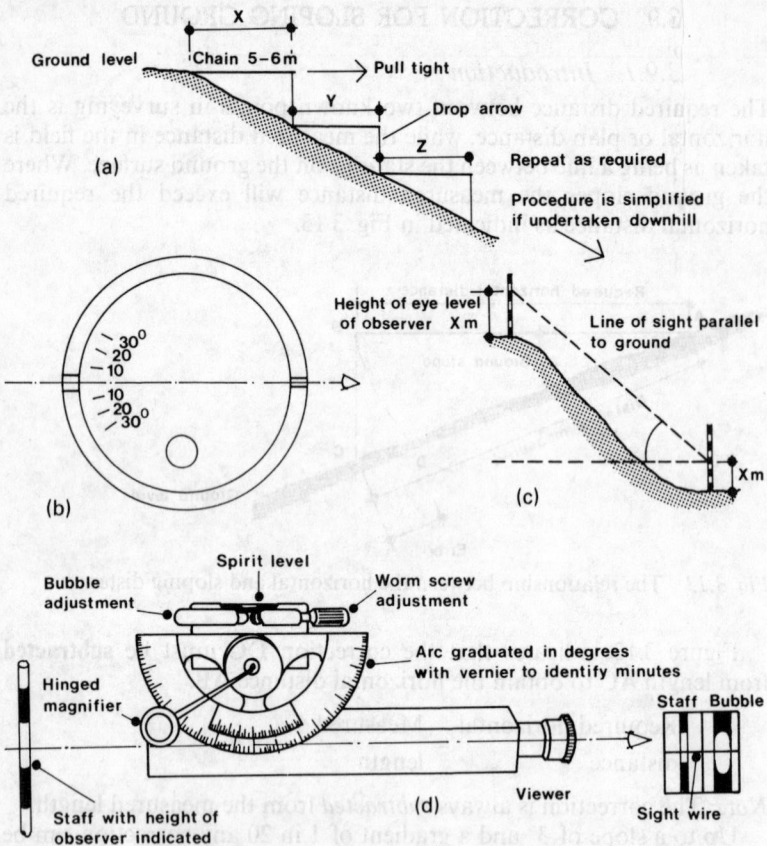

Fig. 3.14 Instruments for ground slope correction. (a) stepping with a chain (b) Watkins clinometer (c) application of the clinometer (d) the Abney level

if the point observed is at a position higher than A then the instrument will be tilted upwards. As the scale is freely suspended its position in relation to the horizontal will not change and the angle of elevation (**red**) or depression (**black**) may be observed from a fixed point on the clinometer when the target is bisected.

The Abney level. This instrument may be used as either a hand-held level or for the measurement of vertical angles. It consists of a sight tube having a draw tube extension to which is attached a graduated arc. An index arm pivoted at the centre of the arc carries a small bubble, whose axis is central to the centre line of the arm. As the bubble is tilted the index arm moves over the graduated arc. The position of the bubble can be observed through the mirror when sighting through the instrument by means of an inclined mirror. (See Fig. 3.14(d).)

Obstacles to field measurement

When using the instrument a sight is taken on to a point at the same height above ground level as the surveyor's eye and the angle on the scale noted.

To measure the angle of slope, the procedure is as follows.

(a) sight on to the work;
(b) bring the bubble into the field of view by means of the milled head;
(c) adjust the bubble by means of the wormscrew so that the bubble is bisected by the sighting wire on the target;
(d) record the angle from the vernier by means of the magnifier.

3.10 OBSTACLES TO FIELD MEASUREMENT

Obstacles may prevent conventional field measurement operations due to one of the following reasons:

3.10.1 Obstacles which prevent measurement but not vision

Figure 3.15(a) outlines a situation where measurement is prevented due to the location of the pond and the following steps may be undertaken:

(i) AB is the survey line over the pond;
(ii) set out perpendiculars EF and CD, such that EF = CD;
(iii) FD may then be measured in the conventional way;
(iv) where reasonable accuracy is required the same procedure may be undertaken on the opposite side and GH measured.

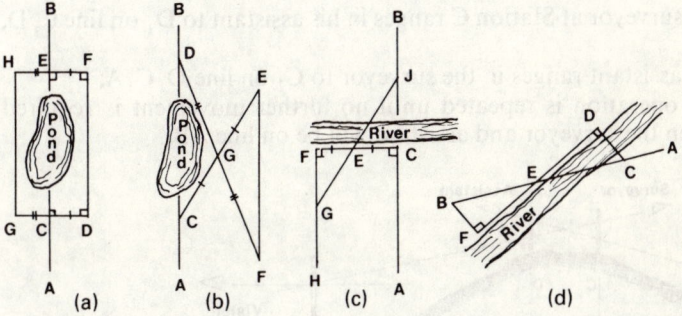

Fig. 3.15 Obstacles which prevent measurement but not vision (a) a pond (b) a pond with space (c) a river at right angles to the chain line (d) a river oblique to the chain line

Where space permits, around the obstacle the procedure illustrated in Figure 3.15(b) may be adopted having the advantage that no angles are set out:

(i) set out CE of any length such that CG = GE;
(ii) set out DF such that DG = GF;
(iii) EF = DC.

The chain survey

A further feature in this class of obstacles is a river where the chain line may intersect at right angles or obliquely.

At right angles. Figure 3.15(c) illustrates this situation.

(i) if AB is the survey line over the river;
(ii) range in Stations C and J on opposite banks of the river;
(iii) set out FC at right angles to the chain line AB;
(iv) locate Station E at the mid-point of line FC;
(v) set out FH at right angles to FC;
(vi) range in Station G on lines FH and JE extended;
(vii) FG = CJ.

Obliquely. Figure 3.15(d) shows the second case.

(i) set out line FED along the river bank intersecting the chain line at E;
(ii) set out FE = ED;
(iii) set out EFG = EDC = 90°;
(iv) GE = EC.

3.10.2 Obstacles to the line of sight but not measurement

The following procedure may be adopted when it is required to chain over a hill:

(i) the surveyor and his assistant place themselves with ranging poles at C and D respectively (Fig. 3.16) so that each can see the other three poles in addition to his own;
(ii) the surveyor at Station C ranges in his assistant to D_1 on line C, D, B;
(iii) the assistant ranges in the surveyor to C_1 on line $D_1 C_1 A$;
(iv) the operation is repeated until no further movement is required when the surveyor and assistant will be on line AB.

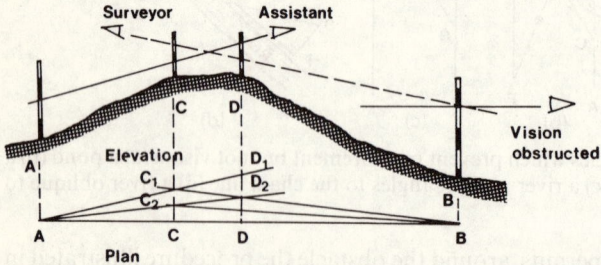

Fig. 3.16 Obstacles to the line of sight but not measurement

This procedure is often called **ranging by trial** and may be used where two stations have been located but fog restricts complete vision on returning to the survey.

3.10.3 Obstacles which prevent both measurement and vision

Where it is required to extend a chain line AD past such an obstacle the following procedure may be adopted:

(a) set out perpendiculars of equal length at C and D such that C and H are beyond the obstacle;
(b) range in J, K on line GH extended, past the obstacle;
(c) set out perpendiculars at J and K as indicated in Fig. 3.17;
(d) locate Stations E and F such that GC = HD = JE = KF;
(e) EF will then lie on line AD extended.

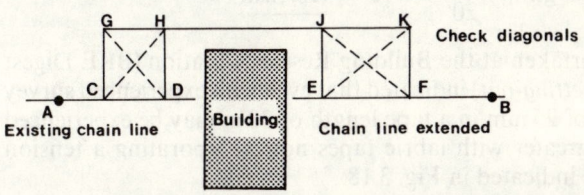

Fig. 3.17 Obstacles which prevent measurement and vision

3.10.4 Survey of a long narrow area

The technique used to overcome such problems is often termed chain traversing and involves the running of consecutive survey lines to follow the area to be surveyed with triangles constructed at the junction between the survey lines. The traverse should be self checking with bearings taken to prominent features.

3.11 ACCURACY OF THE CHAIN SURVEY

The accuracy of a chain survey will be in the order of 1/500 although it can range from 1/100 up to 1/1000 depending on the procedures adopted and the equipment used. The procedure adopted should be capable of checking the following types of error:

(a) omitting to book a chain line;
(b) misreading a chain length;
(c) incorrect booking;
(d) mistaking dimensions which are called out;
(e) error in alignment.

A careful check of the equipment prior to undertaking a chain survey and care of equipment during the work will help to reduce the following:

(a) chains short due to bent links;
(b) slopes over 7° not corrected;
(c) links filled with mud;
(d) stretching of tapes;
(e) bellying of tape.

The chain survey

If a chain has a large error it should be overhauled and brought back to the correct length. Alternatively, it can be used in its incorrect state and each chain line corrected by the following:

$$\text{Correct length of line} = \frac{\text{length of chain used}}{\text{length of standard}} \times \text{measured length}$$

Example 3.2

A line is measured with a chain which is believed to be 20 m long, giving a total chain length of 412 m. On checking the chain was found to measure 20.2 m.

$$\text{Correct length} = \frac{20.2}{20} \times 412 = \underline{416.12\,\text{m}}$$

Research undertaken at the Building Research Station (BRE Digest 234: *Accuracy in setting-out*) indicated that even with experienced survey personnel errors of 25 mm in a tape length of 30 m may be experienced and this will be greater with fabric tapes not incorporating a tension control handle as indicated in Fig. 3.18.

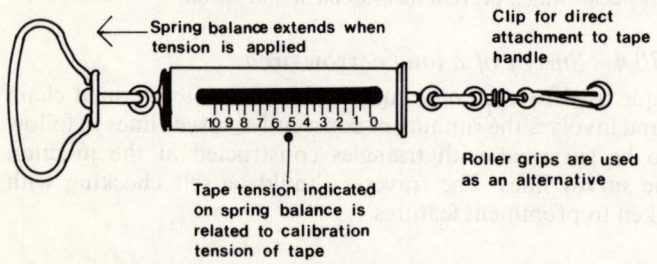

Fig. 3.18 Spring balance tension handle

Example 3.3

A survey line was found to measure 240 m at a temperature of 17 °C. The line had been measured with a 100 m steel band whose true length was 100.005 m at 20 °C. If the coefficient of expansion of the band is 0.000 012 per 1 °C, what is the true length of the survey line?

The temperature at which the band measures 100 m = t.

$$100.005 - 100 = 100 \times 0.000\,012 \times t.$$
$$t = 5/1.2$$
$$t = 4.167°$$

tape is 100 m at 20 − 4.167 = 15.833 °C.

$$\text{Change in length} = 240 \times 0.000\,012 \times (17 - 15.83)$$
$$= \underline{0.003\,4\,\text{m}}$$

$$\text{Correction in length} = 240\,\text{m} + 0.003\,4 = \underline{240.003\,4\,\text{m}}$$

Plotting the survey

This error will only be significant in very hot or cold and an approximate correction is to adjust the linear measurement by 1 mm in 10 m for every 10 °C difference between air and calibration temperature.

Table 3.2 Errors in linear measurement

Cause of error	Error	Comment
Reading	Seldom significant	Use good quality tapes with fine graduations
Calibration	Unlikely to exceed 3 mm	BS 4035 states that graduations should be within ±1.5 mm
Tension	Largest source of error steel tapes calibrated at a tension of 44.5 n. May operate between 9 N and 175 N	Tension control is essential for accurate work. May attach a spring balance to record tension which is related to calibration tension on tape
Tape sag	Causes an over-estimate of the distance. When related to tension errors of 20 mm to +15 mm in 30 m can be experienced	For accurate measurement in catenery, manufacturer's recommended tension and correction factor should be applied
Difference in level	Significant on gentle slopes	Over short distances hold tape horizontal, alternatively apply slope correction
Thermal Movement	Correct where conditions outside calibration temp.	Tapes normally corrected at 20 °C.

3.12 PLOTTING THE SURVEY

Decide on the scale of the plan with regard to the level of detail required and verify that the paper used will accommodate all the survey. Start plotting the survey by drawing the base line of the framework using a straight edge so that all the remaining framework can be plotted. From the base line the remaining stations can be located using beam compasses to identify the remaining stations by the intersection of arcs from two known stations. Before plotting the detail check the accuracy of the framework by measuring all the check lines and compare the framework with the field notes.

The chain survey

Once the framework is considered satisfactory, the detail may be plotted relative to each leg, using a sharp pencil. When plotting the detail, work along one chain line at a time cross-referencing with other chain lines where possible. At this stage an offset scale may be used which is placed along the normal scale, having aligned it on the chain line. This allows the offset scale to slide along the chain line at 90° and detail to be speedily located. Ties may be located by swinging arcs of the correct length from the chain line. Where detail has been located from more than one chain line, the accuracy of the work can be validated.

Once the plotting has been checked to ensure that no errors have occurred the final plan may be produced in ink with a key to identify the symbols, a scale, a north point, border around the sheet and a title block. If required, the work may be colour washed to highlight detail.

3.13 EXERCISES

1. During the course of a chain survey it becomes necessary to set out and measure a long survey line which is obstructed by a large irregular shaped building so that the ranging of the line is not possible because the view is obstructed. Describe in detail, with the aid of sketches, one method by which the direction of the line may be produced after passing the building and also how to ascertain the intercepted distance. An instrument for setting out right angles is not available. (RICS)

2. An area of land is to be developed as a building site and it is necessary to prepare a plan of the area. Describe briefly the method of carrying out the survey of the area using a chain, poles and a linen tape only. Give an example of one page of the fieldbook taking one line of the survey which shows the measurements that would be made to one of the buildings, so that its position may be plotted on the plan. (RICS)

3. Demonstrate the necessity of using well conditioned triangles in chain surveys by calculating the percentage difference between the maximum and minimum areas of the two triangles shown in Fig. 3.19 given that each length is accurate to plus or minus 0.2 per cent. (RICS)

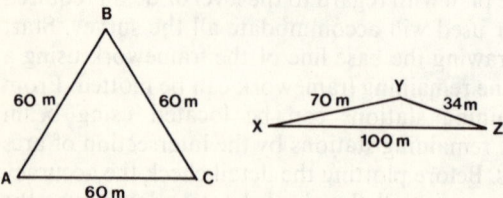

Fig. 3.19 Well conditioned triangles

4 BEARINGS

4.1 TYPES OF BEARING

A bearing allows the direction of an observation to be recorded by means of angular measurement. Bearings for survey work are normally recorded in one of the following formats:

Whole circle bearings (WCB). This method takes North as zero and all bearings are expressed as an angle measured **clockwise** from North (Fig. 4.1). The magnitude of a whole circle bearing may vary between 0 and 360°.

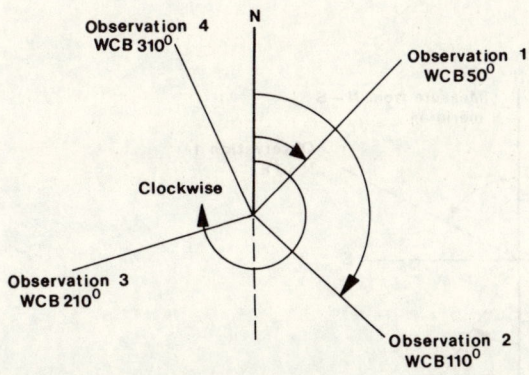

Fig. 4.1 The whole circle bearing

When proceeding from Station A (Fig. 4.2) to the next station at B the angle NAB may be recorded and this is called a **forward bearing**. At Station B angle NBA may be recorded and this is called a **back bearing**.

Bearings

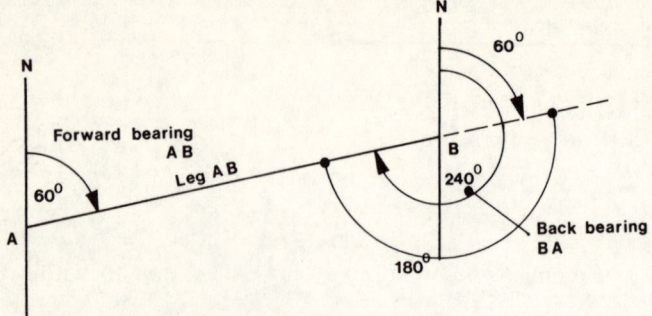

Fig. 4.2 A forward and back bearing

Since angle NAB = Angle NBD
then back bearing − forward bearing = $180°$

The back bearing may be used to check the complementary forward bearing and is found by adding 180° to the forward bearing if it is less than 180° or subtracting 180° if it is greater.

Quadrantal bearings. This method divides the whole circle into four quadrants of 90° and in each quadrant the bearing is always measured from the North–South meridian but may be in a clockwise or anti-clockwise direction (Fig. 4.3).

This angle will never be greater than 90° and must always include North or South first, then East or West to identify the direction of rotation.

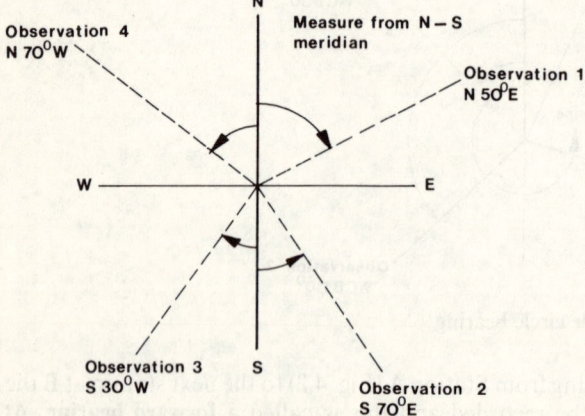

Fig. 4.3 The quadrantal method

4.2 CHOICE OF MERIDIAN

When a bearing is observed it will be related to some fixed reference plane or **meridian** which passes through the point of observation. The meridian used by the Surveyor will depend on the information available and the application of the work and may be one of the following:

4.2.1 The true meridian

This is the meridian of **longitude**, which is a plane passing through the North and South poles on the earth's surface. This meridian can be located by astronomical observations or by calculation to within 6 seconds from the reference grid of an Ordnance Survey map. An observation which is referenced to this meridian is often referred to as **azimuth**.

4.2.2 Magnetic meridian

This is the magnetic North as determined by a magnetic compass free from local attraction. The magnetic meridian is constantly changing since variations in the magnetic field occur due to:

Secular variation. This causes the largest variation and is a slow continuous swing with a cycle of 400 to 500 years. The present movement is approximately 1/2° East every eight years and it should coincide with the true meridian in the year 2055. It is important that maps containing **true** and **magnetic** North include the date when the map was completed and the annual rate of change in order that the present magnetic declination can be calculated. After recording the magnetic bearing of an observation this can be converted to a true bearing by **subtracting** the known declination if the variation is **west** or **adding** the declination if it is **east** of true north, as illustrated below.

Magnetic WCB   235°
True WCB   232°
Magnetic declination 3° West.

Diurnal variation. These are more or less regular variations of 10' to 15' per day about a mean position.

Local attraction. Variations occur due to the presence of magnetic material or electrical interference which causes unpredictable magnetic attraction of the compass needle. It is important that the field procedures adopted when making an observation and the method of booking detects these variations (eg comparison of forward and back bearings), since these may be large and cannot be predicted.

It is because of the complexity of these variations on a magnetic observation that compass survey cannot be very accurate, although for low order work it is only essential to consider secular and **local** variations.

Bearings

4.2.3 Grid meridian

This is derived in the United Kingdom from the National Grid based on a reference meridian 2° West of Greenwich. The meridian is defined by the parallel axes of the grid and the whole circle bearing is the clockwise angle between a grid north line and the station observed. Since the grid lines are parallel, unlike lines of longitude, it is only on the **reference** meridian that **grid North** and **true North** will coincide. East or West of this meridian corrections will be required and this information is provided on Ordnance Survey map sheets, as outlined below:

1. A protractor is used to measure the **grid** bearing between two stations identified on the map (say 234° 10′).
2. The map states that the deviation between grid North and true North is 20° 9′ West, then this correction must be subtracted from the grid bearing.

    eg Grid bearing = 234° 10′

    True bearing = 234° 10′ − 2° 9′

    <u>232° 1′</u>

3. Given the magnetic declination, this bearing can be corrected to provide the magnetic bearing, enabling the station to be located with a compass in the field.

This illustrates that providing the deviation between each meridian is known then the bearing can be adjusted from one reference meridian to another. This is particularly important where bearings are measured from the grid meridian with a protractor since they must be converted to magnetic bearings if the field survey adopts a magnetic compass.

4.2.4 Arbitrary meridian

This can be any carefully chosen reference meridian selected by the surveyor in the field. The reference meridian will be identified by some permanent and clearly defined object such as a church spire which is clearly marked on a map sheet. By taking reference bearings from this meridian and later determining the true or magnetic bearing of the meridian from the information provided by the map sheet, all the field observations can be corrected.

4.3 THE PRISMATIC COMPASS

The prismatic compass (Fig. 4.4) has long been used for military applications and may be used for a limited survey where it will be found to have the following advantages:

(a) the equipment is light and easy to carry;
(b) bearings are quickly observed and simple to check, requiring a minimum of calculation;

The prismatic compass

(c) each bearing is observed independently so that errors may be confined to that section of the survey;
(d) the check lines demanded by a chain survey are not required.

With regard to the above advantages, the following limitations must be considered:

1. The accuracy of the compass may be limited to $\frac{1}{2}°$.
2. Local attraction can seriously affect the observations.

The compass consists of a compass box having a glass top and screw on metal lid, capable of being screwed on to a tripod for location over a station. The case contains an aluminium ring attached to a light magnetic needle which rests on a pivot attached to the centre of the base. The metal ring, called a compass card, is graduated from 0° to 360° in a clockwise direction from the South Pole to provide a direct reading. A damper is fitted which allows the needle to be activated when a reading is taken and reduces damage during transit.

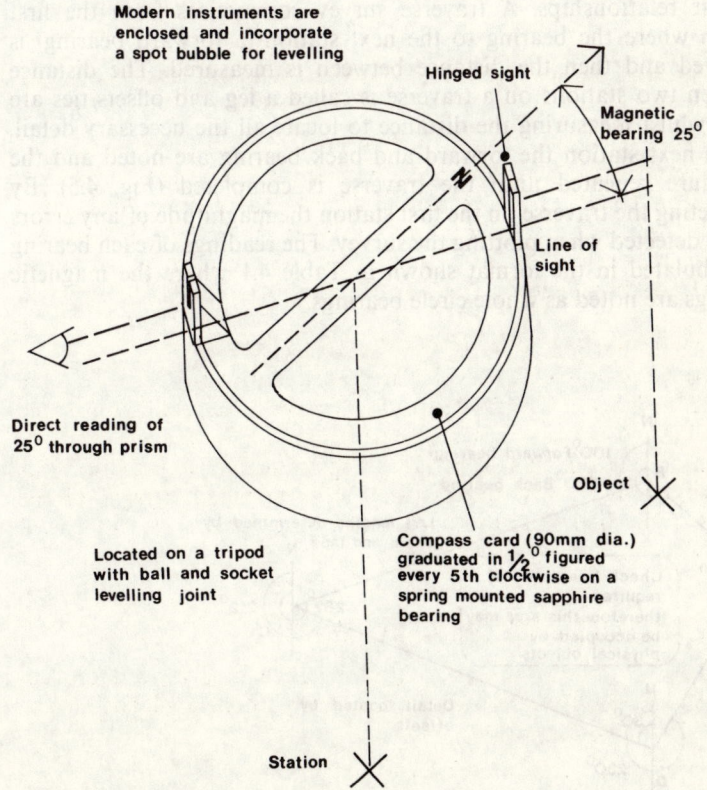

Fig. 4.4 The prismatic compass

Bearings

When making an observation the compass is located centrally over the station and the object, the bearing of which can be sighted by means of a fine slit on top of a hinged prism placed over the compass and a sighting wire on the opposite side of the frame. Since the compass card is graduated from the South Pole a direct reading may be obtained through the hinged prism. The scale is graduated to $\frac{1}{2}°$, although a closer approximation may be made. Some prismatic compasses are damped by being filled with a viscous liquid which improves the speed of an observation.

4.4 THE COMPASS TRAVERSE

4.4.1 Definition

The traverse survey is a method of locating a series of features by measuring the distance between the selected stations and their respective angular relationships. A traverse survey commences from the first station where the bearing to the next station (a forward bearing) is observed and then the distance between is measured. The distance between two stations on a traverse is called a **leg** and offsets/ties are taken whilst measuring the distance to locate all the necessary detail. At the next station the forward and back bearing are noted and the procedure repeated until the traverse is completed (Fig. 4.5). By completing the traverse on the first station the magnitude of any errors can be detected when plotting the survey. The readings of each bearing are tabulated in the format shown in Table 4.1 where the magnetic bearings are noted as whole circle bearings.

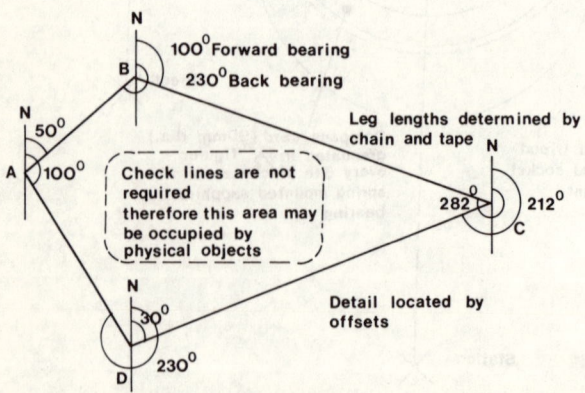

Fig. 4.5 The compass traverse

4.4.2 Correction for errors

This stage should be completed in the field as the readings are booked, in order that gross errors can be identified and new readings taken if required. The correction procedure should ensure that:

(a) the same correction is applied to each bearing observed from any one station;
(b) the difference between the forward and back bearing of each leg is equal to 180°.

Table 4.1

Line	Observed bearing	Difference	Comment
AB	50	180	Forward bearing
BA	230		Back bearing
BC	100	182	Forward bearing
CB	282		Back bearing
CD	212	182	Forward bearing
DC	30		Back bearing
DA	280	180	Forward bearing
AD	100		Back bearing

It will be seen in Table 4.1 that the difference between each forward and back bearing is 180°, except for legs BC and CD. Since AB and AD are correct then Stations AB and D must be free from local attraction. The results indicate that legs BC and CD contain an error and this must have occurred at Station C, since the remaining stations are correct. The attraction at Station C has caused an error of 2° to be identified and this must be subtracted from both the back bearing to Station B and the forward bearing to Station D, giving the corrected readings shown in Table 4.2. If no two readings differ by 180° the corrections should be made from the mean value of the bearings of the line which has the least discrepancy between the **backsight** and **foresight**.

4.4.3 Plotting the survey

Once the results have been corrected the survey may be plotted in pencil to allow corrections and amendments to be undertaken, in the following order:

(a) sketch the outline of the traverse to scale in order to evaluate its approximate position on the paper;
(b) start at the first station by marking its position on the paper and indicate the magnetic meridian;
(c) measure the bearing of the first leg and mark its direction, using a protractor to measure the angle.

Bearings

Table 4.2 Bearings of the compass traverse

Line	Observed bearing	Difference	Correction	Corrected bearing
AB	50°		0°	50°
		180°		
BA	230°		0°	230°
BC	100°		0°	100°
		182°		
CB	282°		−2°	280°
CD	212°		−2°	210°
		182°		
DC	30°		0°	30°
DA	280°		0°	280°
		180°		
AD	100°		0°	100°

(d) measure the correct length of the first leg and locate the second station;
(e) repeat the procedure until the traverse is complete when the detail may be located by means of offsets and ties from each traverse leg.

4.4.4 Adjusting the traverse

It will usually be found that the traverse will not close at Station A, the first station (Fig. 4.6), due to mis-measurement in the field and drawing inaccuracies. The errors are corrected graphically by Bowditch's rule, since it adjusts the angular relationships. This is where relatively large errors may occur with compass work. The rule states that:

$$\text{Correction} = \frac{\text{Length of traverse leg}}{\text{Total length of traverse}} \times \text{closing error}$$

The correction is applied as follows:
Draw a line and plot AB, BC, CD, DA' to any convenient scale.
At A' draw a line parallel to the closing error (indicated in Fig. 4.6) and equal in length to give A.
Join A'A and draw Bb, Cc, Dd parallel to A'A.
Draw Bb, Cc, Dd, at each respective station and locate b, c, d on the survey.
Join A, b, c, d, to give the correct traverse.
The error in a compass traverse is approximately 1/400 and should only be corrected if it is within these limits and:

$$\frac{\text{Closing error}}{\text{Length of traverse}} = \frac{1}{400}$$

If the survey is outside these limits the work should be checked and possibly repeated.

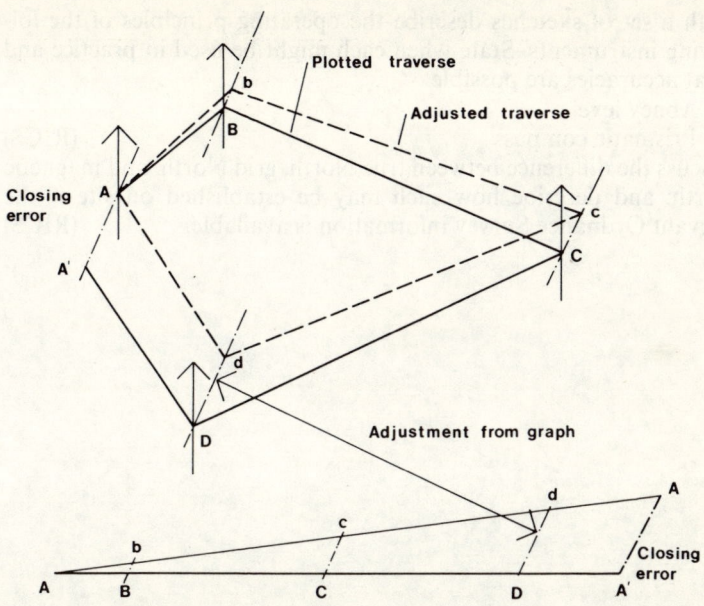
Fig. 4.6 Correcting the traverse

4.5 EXERCISES

1. (a) Make an annotated sketch of a prismatic compass.
   (b) The details of a compass traverse are listed below. Identify those stations where there is local attraction, adjust the bearings, and tabulate the corrected true forward bearings of the traverse if magnetic declination was 5° 30′ W.

Line	Forward bearing	Back bearing
AB	38°	218°
BC	103°	283°
CD	74°	257°
DE	153°	330°
EF	227°	47°
FG	272°	90°
GH	231°	54° 30′
HA	330° 30′	149°

(RICS)

Bearings

2. With a set of sketches describe the operating principles of the following instruments. State when each might be used in practice and what accuracies are possible
   (a) Abney level
   (b) Prismatic compass (RICS)
3. Discuss the difference between true North, grid North, and magnetic North: and describe how each may be established on site if the relevant Ordnance Survey information is available. (RICS)

5 THE THEODOLITE

5.1 FUNCTION OF THE THEODOLITE

Theodolites are designed to measure both vertical and horizontal angles by means of two graduated circles placed at right angles to each other. The instrument is used by surveyors for the measurement of angular work because of its versatility and reliability. When used correctly, it has a wide range of applications, including:

(a) angular measurement between two or more points, together with the inclination of these points from the horizontal;
(b) setting out steep gradients;
(c) the location of co-ordinated setting out points, grids, etc;
(d) optical plumbing of buildings.

5.2 THE PRINCIPLES OF THE INSTRUMENT

Figure 5.1 indicates the basic configuration of the theodolite which consists of a telescope to increase the range of vision and two plates or circles with scales indicating 0° to 360°, at 90° to each other. These plates allow the following angles to be measured:

The horizontal angle. This is measured on the horizontal circle which rotates about the **vertical axis** of the instrument and consists of two plates on many instruments, the lower circle containing graduations and the upper circle an index mark. Each plate has its own clamp screw for

The theodolite

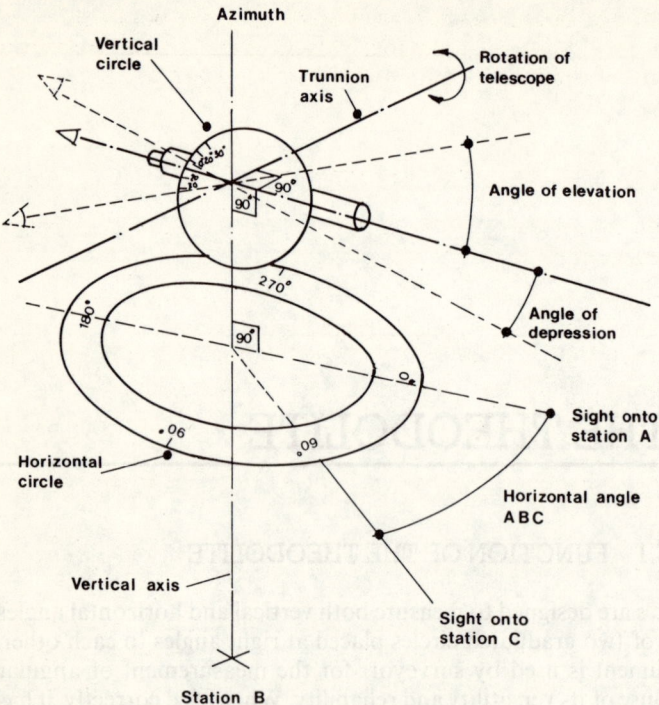

Fig. 5.1 Angular measurement with the theodolite

locking and a slow motion or tangent screw (Fig. 5.2) for fine adjustment. The clamp screws are arranged so that the plate may revolve independently, or:

1. The lower plate may be locked and the upper plate is free to revolve,
2. The upper and lower plate are locked and revolve together,
3. Both plates are locked and no movement is possible.

The vertical angle. This is measured on the vertical circle which rotates about the horizontal **trunnion axis** at 90° to the **vertical axis** and about which the telescope rotates. In its normal position the vertical circle will be on the left-hand side of the telescope and this is called the **face left position**.

When a vertical angle is observed above the line of collimation this is called an angle of **elevation**, whilst an angle below the line of collimation is called an angle of **depression**. Some instruments incorporate a vertical circle compensator which automatically indexes the vertical circle when the instrument is minimally out of level.

5.3 THE AXES OF THE INSTRUMENT

Figure 5.1 indicates the relationship between the three principal axes of the instruments and shows that:

(a) the line of collimation is at right angles to the trunnion axis;
(b) the trunnion axis is at right angles to the vertical axis;
(c) when the line of collimation is horizontal it should be at right angles to both the vertical axis and the trunnion axis.

5.4 THE COMPONENTS OF THE INSTRUMENT

The theodolite (Fig. 5.2) consists of two basic parts and these are common to all modern instruments:

1. *The tribrach.* This is the base of the instrument and incorporates three foot screws for levelling the instrument once it has been located on the tripod. Many modern theodolites have been designed so that a range of instruments can be located over a station, eg targets, electronic distance measuring equipment and subtense bars. This is done by clipping the instrument on to a common tribrach, once the tripod and tribrach have been located over the station. This facility reduces the number of times the equipment must be centred over the station and improves the accuracy of the work. Some tribrachs are detachable and have provision for a ball-centring device to be fixed, by means of a central thread. This enables the theodolite to be mounted on permanent pillars fitted with centring sockets.

2. *The alidade.* This term applies to the horizontal and vertical plates, instrument case and all the moving parts of the instrument, including:

(a) the plate bubble on top of the upper circle which is used for levelling the instrument;
(b) the clip or slow motion screws and plate clamps sighting the instrument on to the reference objects;
(c) the altitude bubble by which the angle of elevation or depression is related to the vertical circle;
(d) an optical plummet which rotates with the alidade to locate the vertical axis over the station;
(e) the telescope supports;
(f) the telescope.

The theodolite

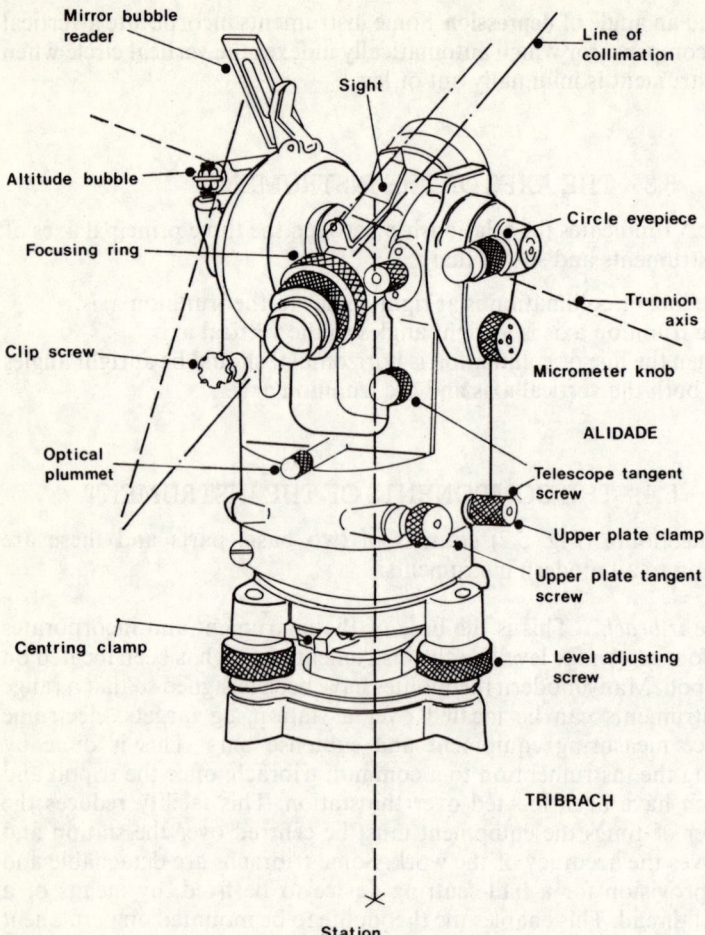

Fig. 5.2 The microptic theodolite (courtesy Hall and Watts Ltd)

5.5 THE SIMPLE TELESCOPE

5.5.1 Description

The telescope incorporated in modern surveying equipment is internal focusing with the eyepiece and object lens mounted at opposite ends of a tube of fixed length (Fig. 5.3). A moving concave lens mounted on a rack and pinion is placed between them to focus the image on the diaphragm. The diaphragm incorporates a glass plate (reticule) on which are engraved the vertical axis and the line of collimation. The intersection of these two axes is used to sight the instrument on to a station and the vertical reticule may also be used by the surveyor to verify that his assistant is holding the staff vertical. The reticule plate is held inside the telescope by means of four adjusting screws which enables:

(a) the vertical axis to be adjusted;
(b) the horizontal axis to be adjusted.

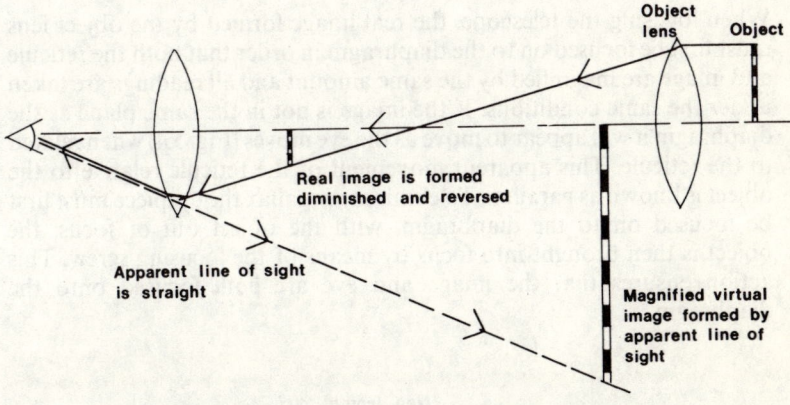

Fig. 5.3 Principles of magnification

This adjustment ensures that the intersection of the reticule coincides with the focal centre of the object lens and eye piece.

5.5.2 *Advantages*
The advantages of the telescope are that it is;

1. Robust in use.
2. Compact to handle.
3. Overcomes the problems of droop experienced with external focusing telescopes.
4. Optical properties are fully utilised (see section 9.2.2, Stadia tacheometry).

5.5.3 *Disadvantages*
The disadvantages are that:

1. The extra lens absorbs some light, although modern lenses are 'bloomed' or tinted to reduce diffusion at the air/glass boundary.
2. The image is reversed, although this can be prevented by the insertion of a further lens within the telescope. The disadvantage of this is that there may be a loss of definition.
3. Spherical aberration, due to the light rays on the outer edge of the curved lens being refracted more than those at the centre, will prevent

accurate focusing. This is reduced in modern telescopes where two lenses are used at either end of the eyepiece.
4. Chromatic aberration is the dispersion of white light into its different wavelengths at the glass/air boundary. This problem is reduced by laminating lenses of different glass to form the object lens.

5.5.4 Parallax

When focusing the telescope, the real image formed by the object lens must first be focused on to the diaphragm in order that both the reticule and image are magnified by the same amount and all readings are taken under the same conditions. If the image is not in the same plane as the diaphragm it will appear to move as the eye moves (Fig. 5.4) when related to the reticule. This apparent movement of the reticule relative to the object is known as **parallax**. To eliminate parallax the eyepiece must first be focused on to the diaphragm, with the object out of focus; the object is then brought into focus by means of the focusing screw. This action ensures that the image and eye are **both** focused onto the diaphragm.

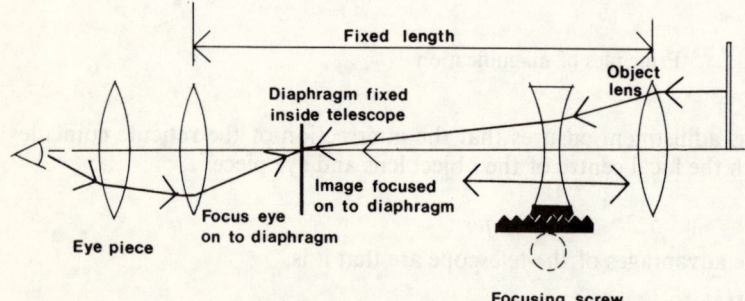

Fig. 5.4 The arrangement of the internal focusing telescope

To check that the setting is correct the telescope should be focused on to a distant object and the eye moved up and down. The distant object and reticule should appear together. If there is any relative movement the adjustment is incorrect and the focusing procedure must be repeated.

5.6 TYPES OF THEODOLITE

Modern instruments are of the transit type, since it is possible to rotate the telescope through 180° about the trunnion axis to improve the accuracy of the instrument. A variety of arrangements may be found with this facility although they can be related to two groups.

Types of theodolite

5.6.1 The vernier theodolite

Description. This instrument has metal vertical and horizontal circles with the scale graduations engraved on both the lower plate and vertical circle. Each circle contains a vernier to allow fractions of the main scale to be accurately measured and a magnifier for accurate reading. The plates are enclosed by a metal case which incorporates a viewer over each vernier scale. Two diametrically opposite verniers 180° apart are usually provided on each scale to increase the accuracy of reading.

Reading the vernier scale. The main scale in Fig. 5.5 is divided into degrees, with each degree divided into three divisions representing 20 minutes of a degree. The vernier is graduated in minutes, each being divided into three divisions representing 20 seconds. This arrangement provides for an accuracy of 20 seconds with each reading. A reading may be obtained by following the procedure below:

(a) locate the index mark and read the scale on the main circle to the nearest whole 20 minute division before the index mark. If the index mark coincides with this division then this will be the reading; if not:
(b) identify a division on the vernier scale which coincides with a division on the main scale;
(c) note the reading of this division on the vernier scale which will be accurate to 20 seconds;
(d) add the vernier reading to the main scale reading to obtain the total reading.

A magnifier is provided to identify the divisions which coincide, when agreement cannot be reached the mean of the divisions in dispute should be taken.

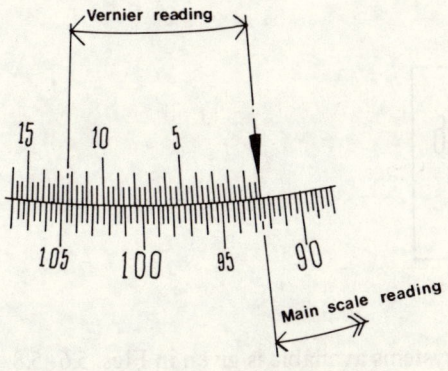

Main scale reading = 92° 20′ 00″
Vernier reading = 00° 12′ 20″
Angle = 92° 32′ 20″

Fig. 5.5 Reading a vernier scale

The theodolite

The accuracy of this theodolite will depend on the diameter of the plate and the fineness of the divisions. It is usually specified according to the finest reading that may be obtained, eg 20 seconds of angle.

5.6.2 The optical systems theodolite

Description. The modern theodolite usually incorporates a glass horizontal and vertical circle having black graduations superimposed on the plate rather than the traditional vernier. Since it is possible to etch finer lines on glass, the circles are reduced in size and consequently a smaller and lighter instrument is provided. An optical system of glass prisms and lenses are provided to refract the light through the glass circle and present the reading at a microscope viewer containing a fixed index hair. Some instruments have a separate viewer for each circle, while other models have one viewer showing both circles simultaneously or even incorporate a change-over switch to view each in turn. Some instruments present a reading which is the mean of two readings from opposite edges of the circle viewed to eliminate scale errors (Fig. 5.8).

Many modern instruments have an optical plummet on the upper plate consisting of a right-angled prism which enables the observer to look, apparently, downwards, along the vertical axis of the instrument. This facility is used to position the instrument accurately over the station without a plumb bob, which is slow and may be affected by wind speed.

Method of reading. A range of reading systems are available, the simplest having a micrometer screw adjacent to the vernier (Fig. 5.7) to allow the scale reading to be moved until the index hair intersects a division. The total angular reading is provided by the addition of the main scale reading and the vernier reading.

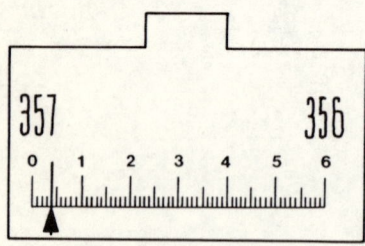

Fig. 5.6 The optical system

A summary of the reading systems available is given in Figs. 5.6–5.8.

The graduations of the horizontal circle (Fig. 5.6) are viewed superimposed by a scale representing one division of the main scale. The magnification normally allows the scale to be read to a decimal point of a minute.

The specification of a theodolite

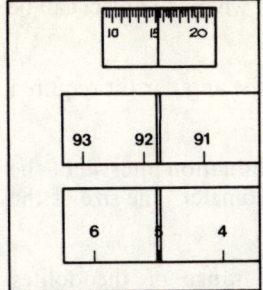

Fig. 5.7 The optical micrometer

In Fig. 5.7 the system works in a similar manner to the parallel plate micrometer, except the line is brought into coincidence with a graduation on the main scale by manual rotation. The amount of displacement required is given by a separate scale viewed through the eyepiece. This type of reading can only be observed at one point on the scale.

Figure 5.8 is similar to the above except the optical system views the diametrically opposite graduations of the circle at the same time. This method reduces the effect of any eccentricity by means of a single reading as opposed to the two readings of the vernier instrument.

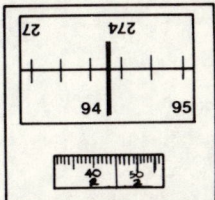

Fig. 5.8 Double reading optical micrometer

5.7 SPECIFICATION OF THE THEODOLITE

The theodolite selected for a particular application will depend on the desired accuracy (section 1.4) of the survey and the results demanded will determine what reading accuracy will be required. To achieve the desired reading accuracy, consideration must be given to:

Aperture. This is the clear diameter of the telescope's object lens. The greater the aperture the more light will enter, clearing the vision.

Magnification. This is the apparent magnification of the object when viewed through the telescope.

The field of view. The width of the field of view is stated in degrees or as the linear width of view at a stated distance.

61

The theodolite

Length of focus. This is the shortest distance at which an object can be viewed through the telescope.

Level sensitivity. This is usually expressed as the angular tilt required to move the bubble 2 mm.

Reading capability. This is expressed as the graduation interval of the scale and the estimating capabilities of the micrometer. The size of the circle will also be specified.

Table 5.1 outlines the specifications of a range of theodolites available for the surveyor, depending on the class of accuracy demanded by the survey. A twenty-second instrument is often satisfactory for low order survey work, although where long sights and a high reading accuracy is demanded a one-second instrument may be used.

Figure 5.9 shows the distribution of deviations incurred by engineers undertaking a standard task of setting out an angle of 90° with **different** classes of instrument. (See G. J. B. Dean and A. J. Stevens, 'Accuracy achieved in setting out with the theodolite and surveyor's level on building sites', Building Research Establishment, *Current Paper* 15/77.) The results clearly demonstrate the importance of selecting an instrument which is consistent with the desired survey accuracy.

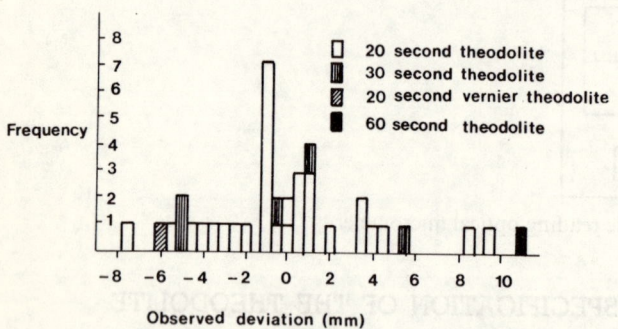

Fig. 5.9 Settling out accuracy of a range of theodolites

5.8 SETTING UP THE INSTRUMENT

5.8.1 Introduction

This procedure is called **temporary adjustment** since it is undertaken each time the instrument is located over a station. There will be no need to remove the instrument from the tripod when moving to new stations since it can be carried on the tripod, providing the central fixing screw is secure and all the instrument's swivel locking screws are secure. The setting up procedures may be considered under the following headings:

Table 5.1 Typical theodolite specifications

Class of theodolite	Telescope aperture	Magnification (X)	Field of view at 1 km	Minimum focusing distance	Circle diameter (mm)	Graduation interval of micrometer	Estimation of micrometer	Mean error of two readings	Centring accuracy	Application
Class I precision survey work	40	30	21 m	2	86	1"	0.1"	±1"	±0.3 mm	Second order triangulatic. Geodetic surveys. Astronomy.
Class II High accuracy survey work	40	30	23 m	1.5	90	1"	0.5"		±0.3	Precise traversing subtense measurement. Astronomical observations.
Class III Engineering and Construction work	36	25	23 m	1.5 m	86	20"	10"	±3"	±0.3 mm	Tacheometric surveys. Setting out roads, drainage. Cadastral surveys.
Class IV Low order detailing	28	20	36 m	1.4 m	68	1'	1'			Photogrammetric ground control survey. Low order detailing of building sites.

5.8.2 Centring

This is undertaken in order to set the **vertical axis** of the instrument directly over the **station**. The surveyor must first check that the tripod head and legs are secure with locking screws that are capable of being tightened. The tripod is first located over the station at a comfortable working height. The tripod plate is approximately levelled by extending or retracting the legs, ensuring that it remains central over the station. The tripod legs should then be firmly trodden into the ground. A tripod base may be used on a smooth hard surface if the feet cannot be secured.

With the tripod firmly located and the upper cap removed ready to receive the theodolite, note how the instrument is placed in its box. Holding the instrument firmly in both hands, attach it to the tripod by means of the tripod fixing screw, take care not to overtighten.

The next stage will depend on the instrument:

Theodolites incorporating a plumb bob. The plumb bob can provide a centring accuracy of ± 5 mm in calm conditions and is attached by means of a bayonet fixing plug in the central fixing screw. Adjust the plumb bob until it is within 20 mm of the ground, noting how far the plumb bob is horizontally from the station. If the whole instrument needs moving then use one tripod leg, moving it up or down or from side to side. Final alignment may be undertaken by releasing the movable head of the instrument and moving the theodolite relative to the tripod until the plumb bob is directly over the station.

Theodolites incorporating an optical plummet. When the instrument includes an optical plummet a centring accuracy of ± 0.5 mm is possible. The eyepiece of the plummet is first adjusted until the cross-hairs are in focus. If the movement required is large then initial centring (Fig. 5.10(1)) may be undertaken by moving the tripod legs. Final centring of the cross-hairs on to the station is undertaken by slackening the tripod fixing screw and **sliding** the instrument horizontally, ensuring that the tribrach is not rotated or the plate bubble is moved off centre. Further adjustment will be required **after** final levelling is completed but this should be accommodated by the movable head (Fig. 5.10(2)).

5.8.3 Levelling

Level the instrument using the three foot screws and bubble on the horizontal plate (Fig. 5.11) as follows:

1. Rotate the telescope until the bubble tube is parallel to two of the foot screws. Turning both foot screws **together** (either **both** inwards or **both** outwards), centre the bubble. The bubble will follow the direction of the **left** thumb when both foot screws are adjusted together.
2. Rotate the telescope through 90° so that the telescope is over the third foot screw. Centre the bubble by adjusting this third foot screw **only** (telescope in position 2).

Setting up the instrument

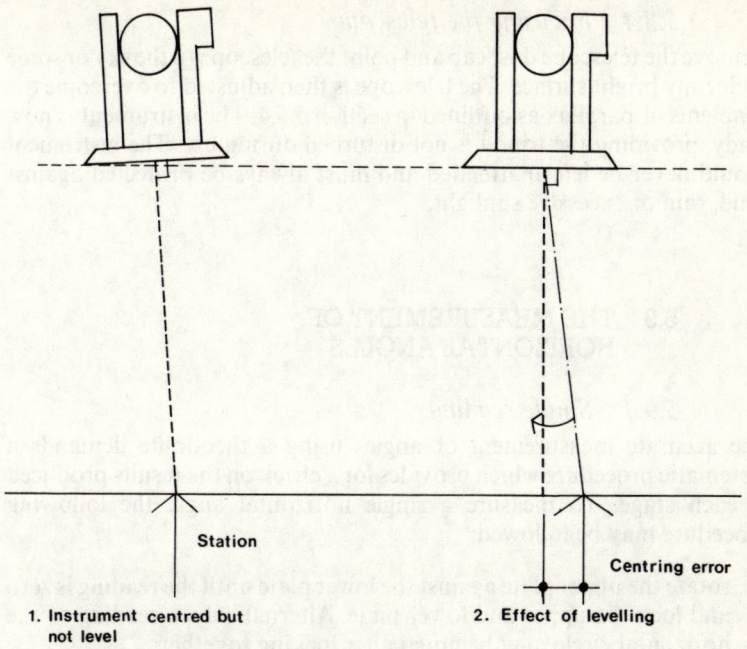

Fig. 5.10 The optical plummet

3. The bubble should now be checked to see that it is central for all positions of the telescope. If further adjustment is required the procedure must be repeated. A shift of one 2 mm division will be satisfactory for most practical purposes. If the bubble will not remain in the centre of its run then the plate bubble may require adjustment (Table 5.2).
4. Final adjustment to the centring may now be undertaken by means of the centring device fitted to the tribrach (Fig. 5.10(2)).

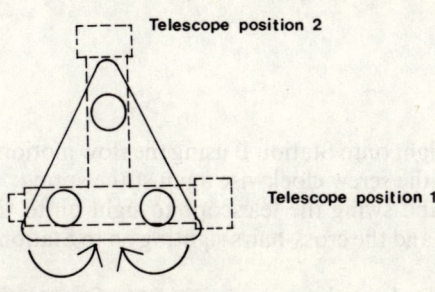

Fig. 5.11 Setting up the instrument

The theodolite

5.8.4 Focusing the telescope

Remove the telescope dust cap and point the telescope to the sky or some uniformly bright surface. The telescope is then adjusted to overcome the problems of parallax as outlined in section 5.5.4. The instrument is now ready, providing the tripod is not disturbed during use. The instrument should never be left unattended and must always be protected against wind, rain or excessive sunlight.

5.9 THE MEASUREMENT OF HORIZONTAL ANGLES

5.9.1 Single reading

The accurate measurement of angles using a theodolite demands a systematic procedure which provides for a check on the results produced at each stage. To measure a single horizontal angle the following procedure may be followed:

(a) rotate the upper plate against the lower plate until the reading is zero and lock the upper and lower plate. Alternatively, a reading of the horizontal circle may be noted after locking together;
(b) with the instrument set up at Station A, sight on to Station B (Fig. 5.12) with the two plates clamped together;

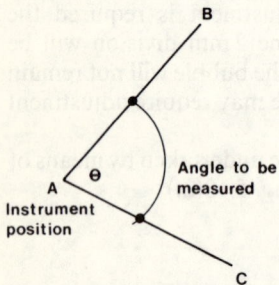

Fig. 5.12 Reading a single angle

(c) clamp the lower plate and sight onto Station B using the slow motion screw with the last turn of the screw clockwise against the spring;
(d) unclamp the upper plate and swing the telescope to sight on to C with the lower plate locked and the cross-hairs sighting on to Station C (Fig. 5.12);
(e) clamp the upper plate and adjust the cross-hairs on to Station C using the upper plate fine adjustment screw, so that the target is accurately bisected;

The measurement of horizontal angles

(f) read the angle recorded on the horizontal plate. If the initial reading (a) was zero then this will be the angle BAC; if the first reading was not zero then the difference between the two readings will give the angle. The above readings should be noted on both verniers of the horizontal plate when a vernier theodolite is used, the mean angle being used for angle BAC.

5.9.2 Repetition

A higher degree of accuracy will be obtained by this method since it reduces any error due to inaccurate graduations of the circle. To measure angle BAC the above procedure is followed and angle BAC obtained. The lower plate clamp is then released and the instrument sighted back on to Station B. With the upper and lower plate clamps locked and the circle reading being equal to angle BAC, the procedure is repeated. The angle may be read several times; then the total angle obtained is divided by the number of readings, to give the mean of angle BAC.

Black station	Station	Forward station	Reading
B	A	C	210° 13' 10"
			60° 26' 25"
			270° 39' 33"

$$\begin{aligned}
\text{First reading} &= 210° \ 13' \ 10'' \\
\text{Second reading} &= 210° \ 13' \ 15'' \\
\text{Third reading} &= \underline{210° \ 13' \ 08''} \\
&\ \ 630° \ 39' \ 33''
\end{aligned}$$

$$\text{Mean reading} = \frac{630° \ 39' \ 33''}{3} = 210° \ 13' \ 11''$$

Note that the calculation of each angle must identify when the scale has exceeded 360°, as was the case with reading 2.
where second reading = (360° 00' 00" − 210° 13' 10") + 60° 26' 25"
= 210° 13' 15"

5.9.3 Reiteration

This procedure may be followed where several angles are measured from one station (Fig. 5.13) when the following procedure may be used:

(a) measure angle AOB and then sight on to Station C;
(b) the reading is then noted to give angle AOC;

The theodolite

(c) the procedure is then repeated until the instrument is again sighted on to the first Station (A);
(d) for improved accuracy the procedure may be repeated but with the instrument set at a different angle when sighting on to Station A;
(e) individual angles may then be obtained by subtraction.

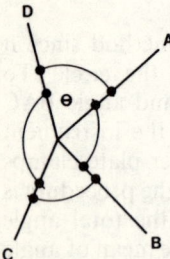

Fig. 5.13 Reading a number of angles

Example of readings

Angle	Reading	Angle	Result
AOB	35° 15′ 10″	AOB =	35° 15′ 10″
AOC	210° 35′ 20″	BOC =	175° 20′ 10″
AOD	330° 40′ 15″	COD =	120° 04′ 55″
		DOA =	29° 19′ 45″

A set of readings as outlined above is termed an **arc** of angles, and where the procedure is repeated after transiting the instrument (Fig. 5.13) the combination of results is called a **round** of angles.

5.9.4 Face left and face right readings

As outlined in section 5.2, it is possible to transit the modern theodolite by revolving the telescope through 180° on the trunnion axis; consequently, the vertical circle may be on the left or right of the telescope when a reading is taken. These are known as face left and face right readings (Fig. 5.14) and will eliminate errors due to imperfect adjustment of the instrument.

Face left is the normal position of the instrument with the sights on top of the telescope. To change face the upper horizontal plate is rotated through 180° and the telescope rotated 180° about the trunnion axis. For extreme accuracy the reading on both verniers should also be noted.

For work demanding a high accuracy it is important that the whole of the horizontal circle is used. In the field this is satisfied by taking face

The booking of observations

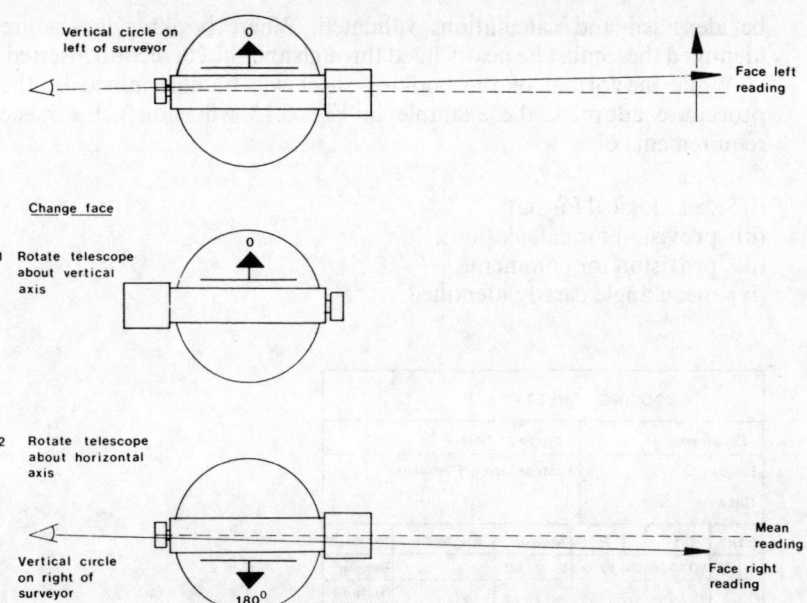

Fig. 5.14 Face left, face right readings

Example of face left and face right readings

	'A' Vernier			'B' Vernier		
Left face on Station A	00	00	00	180	00	20
Left face on Station B	58	24	40	238	25	10
	58	24	40	58	24	50
Right face on Station A	180	00	00	00	00	00
Right face on Station B	238	24	45	58	24	40
	58	24	45	58	24	40

Mean = 58° 24′ 43″

left–face right readings, then repeating the procedure. The mean angle found by this procedure should be lower than the least division of the instrument, while an inspection of the results should identify where a wrong reading has been booked.

5.10 THE BOOKING OF OBSERVATIONS

The booking of observations in the field must allow the angles to be reduced before moving the instrument from the station so that errors can

The theodolite

be identified and calculations validated. Where booking errors are identified these must be neatly lined through and the correction inserted.

While the format of the booking sheet will be determined by the procedure adopted, the example in Fig. 5.15 will satisfy the basic requirements of:

(i) clear logical format;
(ii) provision for calculations;
(iii) provision for comments;
(iv) mean angle clearly identified.

BOOKING SHEET					
Observer		Survey Title			
Booker		Instrument		Remarks	
Date					
Stn		F L	F R	Mean	Angle	Remark
A	B	00 10 00	180 10 00	00 10 00		Ref stn
	C	76 20 10	256 20 05	76 20 08		Station peg

Fig. 5.15 Booking of readings

5.11 MEASUREMENT OF VERTICAL ANGLES

These are measured relative to the vertical plane which may be defined by the altitude bubble after it has first been centred by means of the vertical circle slow motion screw. This is undertaken after levelling the theodolite with the plate bubble and foot screws.

The telescope is first sighted on to the station by means of the horizontal slow motion screw and the vertical bubble centre by means of the clip screws. The vertical circle may be graduated round to 360° with 90° and 270° or 0° and 180° defining the horizontal from which the angle of **elevation** or **depression** will be related.

Many modern instruments incorporate an automatic index which sets the vertical circle horizontal, providing the instrument has been levelled within a given range. A button is usually included to validate the functioning of the index before taking a reading. This causes the vertical circle to swing away from the horizontal and then settle back smoothly. If the graduations appear to jerk back then the instrument is not levelled within the tolerance of the index.

5.12 SOURCES OF ERROR

It is important that the surveyor is aware of errors that can occur when using a theodolite in order that the procedure he adopts in the field can eliminate large errors and identify other errors which may occur. Errors can be identified under the following headings:

5.12.1 Human errors

These may be classified as:

(a) observing wrong station;
(b) reading the angle incorrectly;
(c) using the wrong slow motion screw;
(d) inaccurate centring over the station;
(e) inaccurate target bisection;
(f) inaccurate levelling up;
(g) parallax.

5.12.2 Instrument errors

The following common instrument errors may occur:

(a) imperfect graduations of circle plates;
(b) eccentricity of bearings;
(c) slip on bearings when using slow motion screw;
(d) line of collimation not at right angles to the horizontal axis;
(e) line of collimation not in horizontal axis of telescope;
(f) vertical axis not truly vertical;
(g) trunnion axis not at right angles to the vertical axis.

While errors (a) to (c) require the instrument to be returned to the manufacturer for correction, errors (d) to (g) can be corrected in the field. These corrections are termed permanent adjustments and are summarised in Table 5.2. The above errors will have limited effect on the survey results where a surveyor:

1. Takes face left face right readings to eliminate instrument errors (d) to (g).
2. Reads both verniers and repeats the readings at different positions of the scale.

The importance of adopting the correct operating procedure has been identified in a survey undertaken by the Building Research Station (Fig. 5.16), who observed the error incurred when undertaking the setting out of an angle of 90° with the **same** instrument but adopting **different** reading procedures. (See G. J. B. Dean and A. J. Stevens, 'Accuracy achieved in setting out with the theodolite and surveyor's level on building sites', *Current Paper* 15/77.)

The theodolite

Table 5.2 Permanent adjustments to the theodolite

Error	Effect	Test	Adjustment
Axis of plate bubble not at 90° to the vertical axis	Horizontal angles will not be seriously affected by the inclined plane. Large errors in vertical angles cancelled out by face left/right readings	1. Set instrument level over two foot screws 2. Bring bubble central by foot screws 3. Turn instrument through 180° 4. Bubble should not move off centre	1. Take up half the adjustment on the foot screws 2. Centralise the bubble by the capstan adjusting screw on the bubble 3. Repeat procedure for refinement If bubble moves when upper plate is clamped and lower plate free fault must be rectified by the manufacturer
Line of collimation not at right angles to trunnion axis	Line of sight sweeps out a cone error in horizontal angles at different levels, cancelled out by face left/right readings	1. Sight telescope on to a well-defined object with telescope horizontal and horizontal plates clamped 2. Transit the instrument and mark the position sighted 3. Sight back on to the reference object 4. Transit again and identify the position sighted	1. After undertaking the test two positions will be identified if the instrument is in need of adjustment 2. Move the diaphragm by means of the diaphragm adjusting screws $\frac{1}{4}$ the distance between the outer marks, towards the reference object 3. Repeat the test
Line of collimation perpendicular to the vertical axis	The line of collimation will describe an inclined plane when the telescope is revolved	1. Level the instrument and sight on to an object with an elevation of not less than 45° 2. Depress the telescope and mark a station at ground level	Raise or lower the movable end of the trunnion axis until the sight is midway between the stations as follows: 1. Bring the vertical cross-hair to the mid-point of the stations at ground level by means of the upper plate clamp 2. Sight on to the elevated station

		3. Bisect the target, using the trunnion axis adusting screw
		3. Reverse face and direct on to the reference object
4. If the procedure is repeated the station at ground level will be on the opposite side if the instrument requires adjustment
Adjustment of altitude bubble and vertical circle zero	Vertical circle will not read 0° when altitude bubble is central	1. Level the instrument and sight on to a levelling staff 50 m away
2. Bring the altitude bubble to the centre of its run, using the clip screws. Bringing the vertical circle Vernier to zero with the tangent screw, take a reading and transit the telescope. Repeat the procedure and the second reading should be the same
Set the telescope to the mean reading by means of the vertical circle tangent screw
Arrange the vernier to read zero by means of the clip screw
Adjust the altitude bubble by means of the bubble capstan adjusting screw

The theodolite

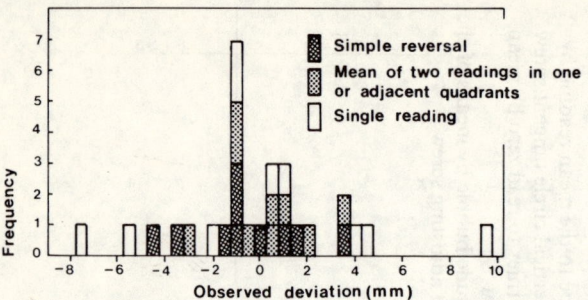

Fig. 5.16 Deviations incurred by engineers when using a 20 second theodolite

5.13 EXERCISES

1. Sketch and describe an internal focusing type of telescope as used in modern surveying instruments. Show the arrangement of the lenses.
   (RICS)
2. Define the following terms as used in connection with a theodolite and its uses:
   (a) Tribrach
   (b) Transiting (RICS)
3. State fully what is meant by:
   (a) Line of collimation
   (b) Backsight
   (c) Parallax (RICS)

6 THE TRAVERSE SURVEY

6.1 ACCURACY OF THE TRAVERSE

A traverse is formed by measuring the angular relationships of adjacent survey stations and the linear distance between them. This information will enable the traverse and related detail to be plotted without the need for tie lines, etc to other stations, as described for the chain survey (Ch. 3).

Since greater accuracy can be achieved in measuring angles with the theodolite than linear measurement, using a chain or tape, the resulting survey plot (Fig. 6.1) will be more accurate. This form of survey is used to locate detail for the lowest order triangulation stations of the Ordnance Survey (section 1.2.2), or at the lowest order of accuracy the principles apply to the compass traverse (section 4.4). This range of accuracy will include the majority of applications required by the professional surveyor.

For work requiring an accuracy greater than the above traverse applications where the error may be less than 1/20 000, stations are located by triangulation (section 1.3.1). In this case a network of connected triangles will be formed from a base line of known length. The lengths of the unknown sides and the relative position of any feature can then be determined with an accuracy up to 1/50 000, once the angular relationships between stations have been determined. The fieldwork for national control surveys may require the absolute position of critical stations to be determined from astronomical observations, but this is outside the range of work of this chapter and the mathematical correction procedures for these applications are covered in more advanced surveying books.

The traverse survey

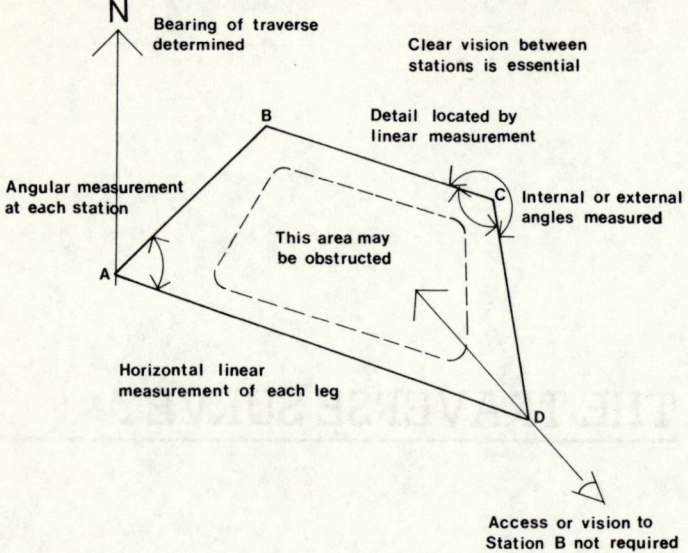

Fig. 6.1 The traverse survey

6.2 FORMS OF TRAVERSE

The traverse used for a particular application will be:

The closed traverse. Figure 6.1 illustrates the principles of the **closed** traverse, which starts and finishes at Station A. Since the traverse forms a **closed** polygon the angular and linear properties can be checked and the accuracy of the survey verified. This principle must be applied to all survey procedures in order to identify any error in the final result and determine whether correction procedures can be applied or the fieldwork repeated. Where the survey does not close but the position of the **first** and **last** station is known from other survey data, then this may be considered as a closed traverse providing the misclosure error can be detected.

The open traverse. A survey where the relationships between the first and last station is not known and the error of misclosure cannot be detected is called an **open traverse**. With this form of traverse it is difficult to identify where errors occur, even when the reference sights and check procedures outlined in Fig. 6.2 are undertaken and should only be used in exceptional circumstances.

Field procedures

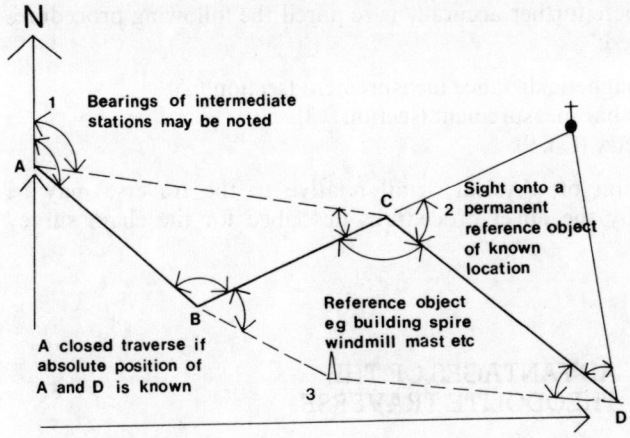

Fig. 6.2 The open traverse

6.3 FIELD PROCEDURES

The survey stations are located by wooden pegs with nails driven in the top or metal studs for concrete, road surfaces, etc. The stations should be located so that the traverse legs are close to the maximum adjacent detail, consistent with the minimum of stations and so placed that each is **visible** from the preceding and following station. Each station should first be 'witnessed' (section 1.5.5) by ties to adjacent permanent points so the location can be validated throughout the survey period.

The survey consists of obtaining angular and linear measurements in order to plot the traverse, and whilst internal or external angles may be booked this description will only consider the measurement of internal angles. It is important to adopt a routine procedure for this work which identifies either internal **or** external angles, to avoid confusion during the subsequent calculations. The survey will proceed from Station A in an anti-clockwise direction with each internal angle obtained by sighting on to the back Station (D) and then on to the forward Station B. The examples used in this section will only indicate one angular reading but this is assumed to be the mean of a minimum of two theodolite readings (section 5.9) or more, where greater accuracy is required.

The first station occupied is normally the reference station from which the remaining stations will be plotted and a bearing is taken of the first leg to one of the meridians outlined in section 4.2 in order to orientate the survey (Fig. 6.1).

The linear measurement of each traverse should be undertaken with a steel band incorporating tension control to relate the accuracy to that provided by the theodolite for angular measurement. The accuracy of the linear measurement should be verified by taping the distance at least

The traverse survey

twice and where further accuracy is required the following procedures may be applied:

(a) electro-magnetic distance measurement (section 6.5).
(b) sub-tense bar measurement (section 9.3).
(c) tacheometry (Ch. 9).

The location of physical detail relative to the traverse may be undertaken by the **offset** procedures described for the chain survey (section 3.6).

6.4 ADVANTAGES OF THE THEODOLITE TRAVERSE

The main survey lines used for the chain survey are influenced by the need to provide well-conditioned triangles, but this does not apply to the traverse survey which provides the following advantages:

(a) the stations are chosen with regard to the physical features to be plotted;
(b) fewer traverse lines have to be set out, therefore the field measurement and consequent plotting is reduced;
(c) sights to intermediate stations on the survey are not required with a closed traverse;
(d) survey stations can be more accurately located;
(e) the traverse survey can be more accurately checked for both field measurement and plotting errors.

6.5 ELECTRO-MAGNETIC DISTANCE MEASUREMENT

The development of micro-circuits has increased the range of feasible applications of this equipment and means the surveyor need no longer rely on the steel tape held in catenary in order to relate the accuracy of linear measurement to the angular measurement accuracy that is possible with a theodolite. The system determines the distance between a transmitter and receiver by recording the transit time of the electromagnetic wave, then converts this into the distance travelled, since the speed of the wave is known.

Modern instruments capable of measuring up to 2–3 km are often used for engineering surveys and setting out since they are accurate, compact and easy to use. They use the infra-red wavelengths and incorporate the transmitter (gallium arsenide (Ga–As) luminescent diode) and receiver in one unit. This unit is located over one station and a

Traverse accuracy and plotting

reflector located over the other station to allow the electro-magnetic wave to return to the transmitter/receiver unit. The development of small micro-processors has enabled the above principles to be included in a small unit capable of being mounted on the theodolite and providing a reading having a root mean square error of ± 5–10 mm. These developments have simplified the operating procedures and many include the following facilities:

(a) an audible signal which remains silent until the beam touches the reflector to aid initial alignment;
(b) automatic slope distance reduction on input of the vertical circle reading of the theodolite which provides a direct display of the **horizontal** distance;
(c) an automatic measuring mode which means the signal strength does not require readjustment if a moving object interrupts the transmitted beam;
(d) automatic atmospheric correction to the display on input of a correction factor related to air temperature and pressure.

Laser systems are also available which are capable of operating up to 50 km, but the presence of mist, fog and rain will considerably reduce the maximum range of operation. The stable light wave improves the accuracy, enabling them to be used for tunnelling and minework.

For distances up to 50 km micro-wave systems are used where the phase difference between the sent and received signal provides a measure of the distance travelled. All of these systems have improved the accuracy of tri-lateration surveys, particularly in inaccessible areas and triangulation networks where linear measurements are required to supplement angular observations.

(Further information may be obtained from *Electro-magnetic Distance Measurement: Aspects of modern land surveying*, by C. D. Burnside, Crosby Lockwood Staples, Granada, St Albans, 1982.)

6.6 TRAVERSE ACCURACY AND PLOTTING

6.6.1 *Field accuracy*

The theodolite traverse will provide a higher degree of accuracy than the chain survey (Ch. 3) or the compass traverse (Ch. 4), although the accuracy achieved will be determined by:

(a) the quality of the instrument used;
(b) the observation and recording procedures followed for both linear and angular measurement;
(c) the procedures used for data adjustment and validation;
(d) the plotting procedures.

The traverse survey

Traverse survey work may be identified at the following levels of accuracy:

1. A precise traverse where angular measurements are taken to 1 inch. Since the respective linear distances may be measured by electronic distance measurement this should provide a closing error no greater than 1/25 000 with an angular error of ($2\sqrt{\text{number}}$ of sides of traverse) seconds.
2. A secondary traverse where angular measurements are taken to 10" with the respective distances taped to an accuracy of 0.01 m. This should provide a closing error no greater than 1/5000 with an angular error of ($30\sqrt{\text{numbers}}$ of sides of traverse) seconds.
3. A tertiary traverse where angular measurements are taken to 300 mm and the ground chaining of distances provides an accuracy of 0.1 m with a similar error in the co-ordinates.

The level of accuracy undertaken in the field will be determined by the objectives of the survey with the reference information specified at a level higher than that demanded by the plotting. For example, a secondary traverse may be related to primary traverse data.

6.6.2 *Plotting accuracy*

The technique used for plotting the survey must be consistent with the accuracy adopted in the field; therefore, the use of a protractor to measure the angular relationships as outlined for the compass traverse (section 4.4) will not be satisfactory. When plotting a theodolite traverse the relative position of one point to another is obtained by **rectangular co-ordinates** rather than polar co-ordinates since:

(a) this provides a level of accuracy consistent with the field work since angular measurements are not undertaken when plotting;
(b) errors can be identified and relevant adjustments made at each stage of the work;
(c) each station is plotted independently, relative to a chosen reference point.

The rectangular co-ordinates adopted consist of the **vertical** and **horizontal** distances of a station relative to a point of origin. The axis running north–south is called the reference meridian and from this reference the co-ordinates are calculated from the following.

Latitude. The is the vertical projection of a line or leg of a traverse and is obtained by the product of the leg length and the **cosine** of the reduced bearing of that leg as illustrated in Fig. 6.3.

Where the latitude is in a northerly direction it is often called a northing and is considered to be a positive value. If it is in a southerly direction it is identified by a negative sign.

Traverse accuracy and plotting

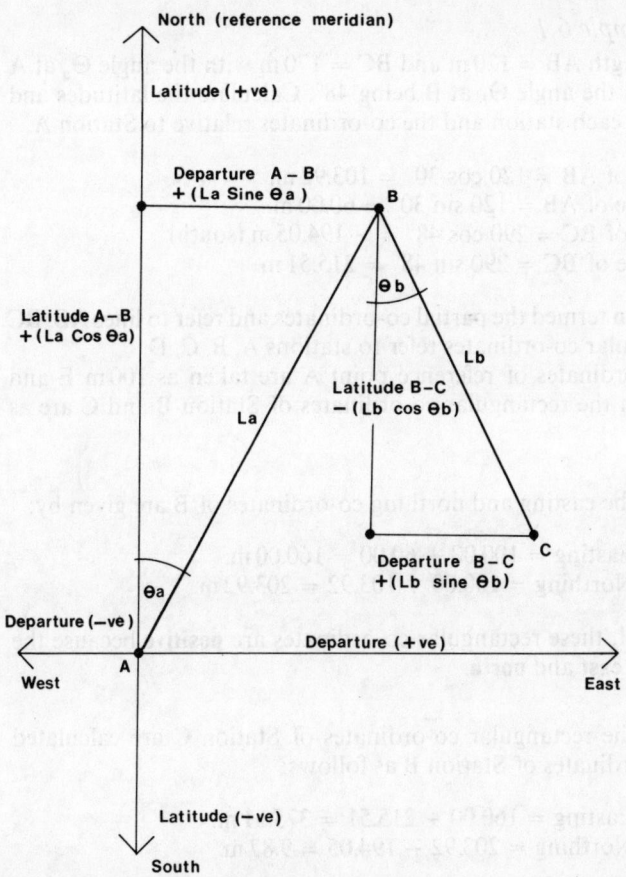

Fig. 6.3 Latitudes and departures

Note that the positive or negative sign is used to identify the **direction** of the latitude as illustrated in Fig. 6.3.

Departures. This is the horizontal projection of a traverse leg and is obtained from the product of the leg length and the **sine** of the reduced bearing of the leg.

Where the departure is in an easterly direction it is often called an easting and is positive. If it is in a westerly direction then it will be negative.

The final co-ordinates of each station relative to Station A in Fig. 6.3 are obtained by the algebraic sum of the respective latitudes and departures since these provide the **partial co-ordinates** of each leg.

The traverse survey

Example 6.1

In Fig. 6.3 length AB = 120 m and BC = 170 m with the angle Θ_a at A being 30° and the angle Θ_b at B being 48°. Calculate the latitudes and departures of each station and the co-ordinates relative to Station A.

(i)   Latitude of AB = 120 cos 30° = 103.92 m.
(ii) Departure of AB = 120 sin 30° = 60.00 m.
(iii) Latitude of BC = 290 cos 48° = −194.05 m (south).
(iv) Departure of BC = 290 sin 48° = 215.51 m.

These are often termed the partial co-ordinates and refer to lines AB, BC while rectangular co-ordinates refer to stations A, B, C, D.

If the co-ordinates of reference point A are taken as 100 m E and 100 m N, then the rectangular co-ordinates of Station B and C are as follows:

Station B.   The easting and northing co-ordinates of B are given by:

The Easting = 100.00 + 60.00 = 160.00 m.
The Northing = 100.00 + 103.92 = 203.92 m.

Note that both these rectangular co-ordinates are **positive** because the directions are **east** and **north**.

Station C.   The rectangular co-ordinates of Station C are calculated from the co-ordinates of Station B as follows:

The Easting = 160.00 + 215.51 = 375.51 m.
The Northing = 203.92 − 194.05 = 9.87 m.

Note that latitude of BC was negative and this is **subtracted** from the co-ordinate of B. The example outlines the principle of the calculations for the rectangular co-ordinates of a theodolite traverse and indicates that these are obtained by adding the partial co-ordinates of each line to the rectangular co-ordinates of the last station, taking account of any negative signs to indicate directions south or west. An outline of the procedure is given in Table 6.1

Where a traverse closes on the first station the sum of the positive and negative latitudes should be zero and the sum of the positive and negative departures should also be zero if the traverse is to close. This provides a check on the accuracy of the fieldwork and calculations prior to plotting the traverse and also identifies the misclosure error.

Example 6.2

A straight road is to be formed between two Stations A and B whose co-ordinates are as follows

Traverse accuracy and plotting

Table 6.1 Outline of procedure for calculating rectangular co-ordinates

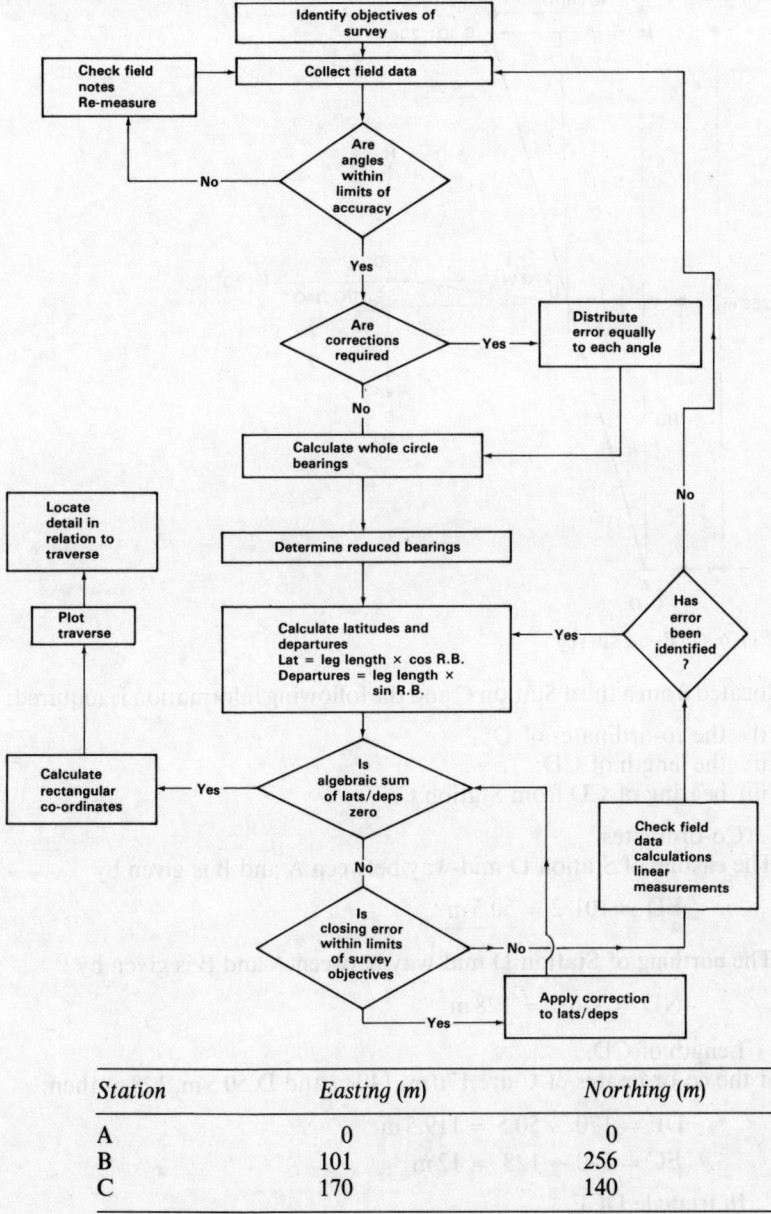

Station	Easting (m)	Northing (m)
A	0	0
B	101	256
C	170	140

It is desired to locate the mid-point D of the road between AB but it is not possible to measure directly along the proposed road. D is to be

83

The traverse survey

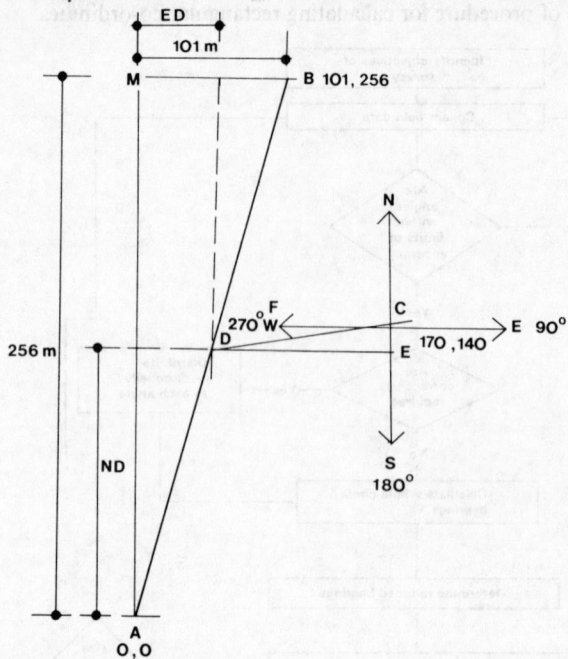

Fig. 6.4 Road survey

located from a third Station C and the following information is required:

(i) the co-ordinates of D;
(ii) the length of CD;
(iii) bearing of CD from Station C.

Co-ordinates.

The **easting** of Station D mid-way between A and B is given by

$$ED = 101/2 = \underline{50.5\,m}$$

The **northing** of Station D mid-way between A and B is given by

$$ND = 256/2 = \underline{128\,m}$$

Length of CD.
If the co-ordinates of C are 170 m, 140 m and D 50.5 m, 128 m then:

$$DE = 170 - 50.5 = 119.5\,m$$
$$EC = 140 - 128 = 12\,m$$

In triangle DCE.

$$CD^2 = 119.5^2 + 12^2$$
$$\underline{CD = 120.1\,m}$$

Bearing of CD.

$$\sin CDE = 12/120.1$$
$$CDE = 5° 44' 4''$$
$$\text{Bearing of CD} = 270° 0' 0'' - 5° 44' 4'' = \underline{264° 15' 56''}$$

Example 6.3

Two points P and Q are inaccessible to each other and it is necessary to find the distance between them. A base line has been set out and observations taken to P and Q by theodolite, from two points A and B.

The angles were measured and the following values obtained (Fig. 6.5):

Angle BAQ 39° 15' 0''
Angle BAP 71° 40' 0''
Angle ABP 45° 20' 0''
Angle ABQ 80° 27' 0''

If the length of line AB is 251.30 m calculate the length of line PQ.

(RICS)

In triangle AQB.

$$AQB = 180° - (39° 15' + 80° 27') = \underline{60° 18'}$$

By sine rule

$$AQ = \frac{251.3}{\sin 60° 18'} \times \sin 80° 27'$$
$$= \underline{285.296 \text{ m}}$$

In triangle APB.

$$APB = 180° - (71° 40' + 45° 20')$$
$$= 63°00'$$

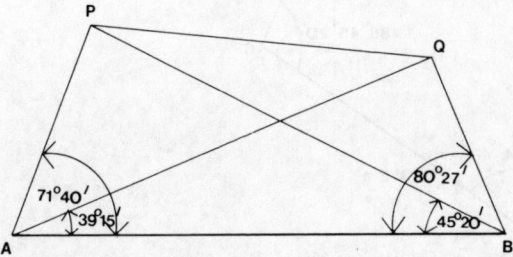

Fig. 6.5 Base line location

The traverse survey

By sine rule

$$AP = \frac{251.3}{\sin 63°} \times \sin 45° 20' = 200.59 \text{ m}$$

In triangle APQ.
(using the cosine rule)

$$PQ^2 = 285.296^2 + 200.59^2$$
$$- (2 \times 200.59 \times 285.296 \cos 32° 25')$$
$$= \underline{158.47 \text{ m}}$$

6.7 EXAMPLE OF A THEODOLITE TRAVERSE CALCULATION

Figure 6.6 illustrates a closed traverse with the mean included angle at each station, and the length of each leg and the magnetic bearing of one leg. The procedure for obtaining the rectangular co-ordinates is outlined in Table 6.1 and includes the steps shown in the following sections.

6.7.1 Field checks of results

Before completing the fieldwork a check of the accuracy of the internal angle readings can be made since the sum of internal angles is $(2n - 4)$ right angles, where n is the number of sides of a closed polygon.

If the external angles had been observed, the sum of the external angles would be $(2n + 4)$ right angles.

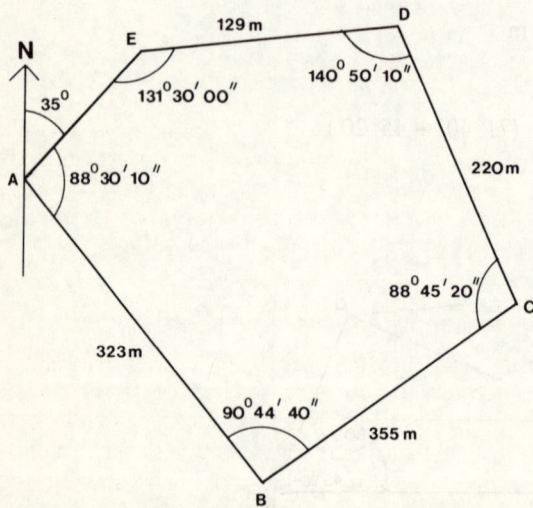

Fig. 6.6 Traverse example

Example of a theodolite traverse calculation

Check of internal angles
Sum of internal angles = 88° 30′ 10″ + 90° 44′ 40″ + 80° 45′ 20″
$$+ \ 140° \ 50′ \ 10″ + 131° \ 30′ \ 00″$$
$$= \underline{540° \ 20′ \ 20″}$$

When $n = 5$: $(2n - 4)$ right angles = 540°
$$\text{Error} = \underline{20′ \ 20″}$$

This level of accuracy would not satisfy the requirements of the secondary traverse precision stated in section 6.6 and would require the fieldwork to be repeated. Assuming that the data relate to a low order tertiary traverse this error may then be acceptable and must be corrected before any calculations are undertaken. Since no indication is given of where the error occurred each angle is corrected by the following amount:

$$\text{Angular adjustment} = \frac{\text{Angular error}}{\text{No. of stations}} = \frac{20′ \ 20″}{5} = 4′ \ 4″$$

This correction will be **subtracted** from each reading to give the following corrected readings:

∠ BAC = 88° 26′ 6″
∠ ABC = 90° 40′ 36″
∠ BCD = 88° 41′ 16″
∠ CDE = 140° 46′ 6″
∠ DEA = 131° 25′ 56″

These values will be used for all subsequent calculations.

Check of linear measurement.   Any errors in linear measurement will be identified later; the only field checks are to measure each leg twice and form a closed traverse. If a gross taping error has occurred on one line only, it will have approximately the same bearing as the closing error and can often be identified.

6.7.2 Calculation of whole circle bearings

By calculating the whole circle bearing (WCB) of each traverse leg relative to the reference meridian it is possible to determine the reduced bearing (RB) for calculating the rectangular co-ordinates. The whole circle bearing was defined in section 4.2 as being the **clockwise** rotation from the north–south meridian to the traverse leg. The calculations commence at the station where a bearing was observed in the field using the corrected angles from section 6.7.1 as follows.

The traverse survey

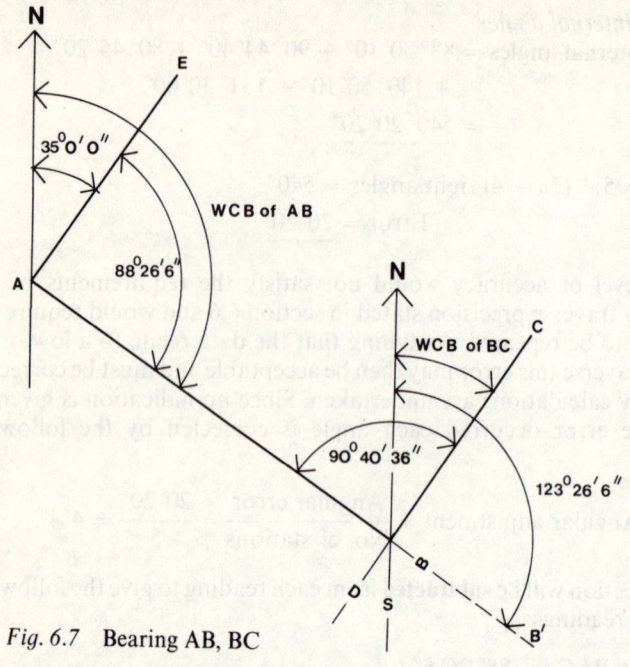

Fig. 6.7 Bearing AB, BC

Bearing AB (obtained from direct measurement) 123° 26′ 6″
Bearing BC
Since the angles NBB′ and NAB are equal (Fig. 6.7),

$$\angle NBC = (\angle ABS + \angle ABC) - 180°\,0'\,0''$$
$$= (WCB\ of\ AB + internal\ angle\ at\ B) - 180°\,0'\,0''$$
$$= \underline{34°\,06'\,42''}$$

Bearing CD
This is indicated in Fig. 6.8 and is equal to:

$$\angle NCD = \angle NCC' + \angle DCB + \angle BCC'$$
$$= (WCB\ of\ BC - internal\ angle\ at\ C) - 180°$$
$$= (88°\,41'\,16'' + 34°\,06'\,42'') + 180°\,0'\,0''$$
$$= \underline{302°\,47'\,58''}$$

Bearing DE
This is indicated in Fig. 6.9 and may be calculated as follows:

$$\angle NDE = (\angle NDD' + \angle D'DF) - 180°\,0'\,0''$$
$$= (WCB\ of\ CD + internal\ angle\ at\ D) - 180°\,0'\,0''$$
$$= (140°\,46'\,6'' + 302°\,47'\,58'') - 180°\,0'\,0''$$
$$= \underline{263°\,34'\,4''}$$

Example of a theodolite traverse calculation

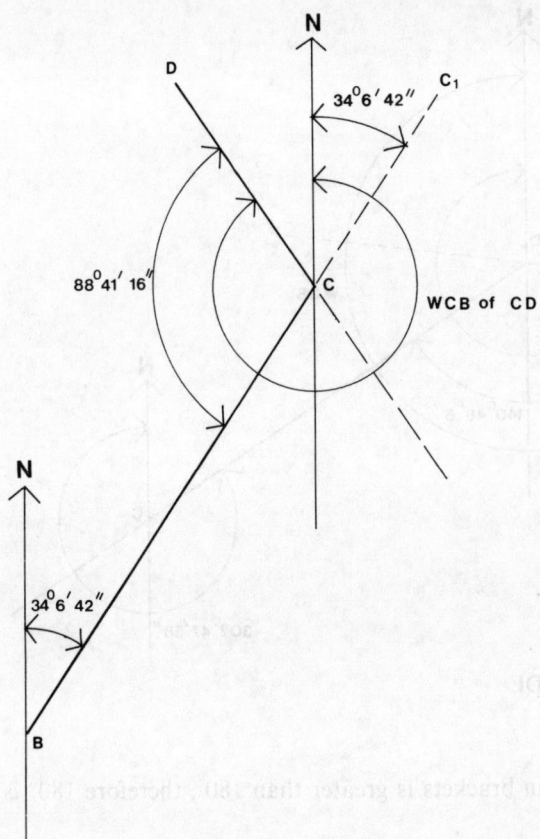

Fig. 6.8 Bearing CD

Note from the above calculations that in each case (other than the first station) that the **included angle is first added** to the previous whole circle bearing and then:

(a) if the sum of the included angle and the previous whole circle bearing is greater than 180° then subtract 180° from the result (leg DE);
(b) if the sum of the included angle and the previous whole circle bearing is less than 180° then 180° is added to the sum (leg BC and CD);
(c) note in the case of leg BC that if the final value exceeds 360° then 360° is finally subtracted.

The whole circle bearing of the first leg is obtained directly from the reference bearing.

The final whole circle bearings can be obtained directly:

Bearing EA
Whole circle bearing = $(131° 25' 56'' + 263° 34' 4'') - 180° 0' 0''$

The traverse survey

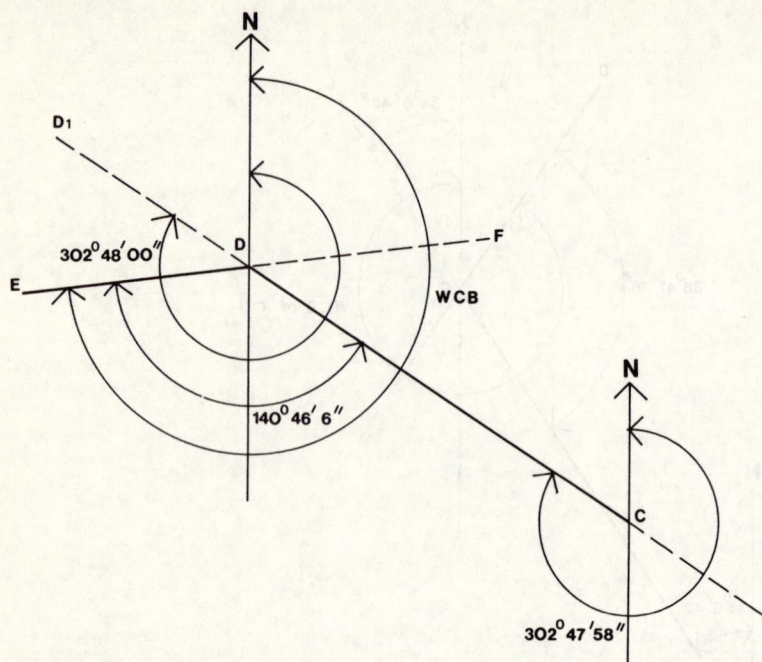

Fig. 6.9 Bearing CD, DE

Note that the value in brackets is greater than 180°, therefore 180° is subtracted.

$$\text{WCB} = 395°\,00'\,00'' - 180°\,00'\,00'$$
$$= 215°\,00'\,00''$$

Bearing AB. This was the first whole circle bearing to be calculated and can be used to check the procedure using the above steps.

$$\text{WCB of AB} = (88°\,26'\,06'' + 215°\,00'\,00'') - 180°\,00'\,00''$$
$$= 123°\,26'\,60''$$

This validates the calculations, as it is equal to the first WCB that was measured directly from the meridian.

6.7.3 Calculations of reduced bearings

Since the whole circle bearing identifies the quadrant of the respective leg, the reduced bearings may be obtained from direct observation (Table 6.2). Note that the directions N.S.E. or W. must always be quoted as described in section 4.1.

Example of a theodolite traverse calculation

Table 6.2 Table of reduced bearings

Leg	Whole circle bearing	Reduced bearing
AB	123° 26′ 06″	S 56° 33′ 54″ E
BC	34° 06′ 42″	N 34° 06′ 42″ E
CD	302° 47′ 58″	N 57° 12′ 02″ W
DE	263° 34′ 04″	S 83° 34′ 04″ W
EA	215° 00′ 00″	S 35° 00′ 00″ W

6.7.4 Calculation of latitudes and departures

The calculations of the vertical projection (latitude) and the horizontal projection (departure) from the meridian may be undertaken once the reduced bearings and length of each leg is known, remembering that:

Latitude = Leg length × cosine of reduced bearing.

Departure = Leg length × sine of reduced bearing.

Note that the positive or negative signs are determined by the reduced bearing.

The latitudes and departures identify the position of a station with respect to the last station and are referenced to the meridian. Since the example is a closed traverse (starts and finishes at Station A), the sum of the latitudes and departures should be zero (Table 6.3).

Table 6.3 Latitudes and departures

Leg	Length (m)	Reduced bearing (RB)	Latitude	Departure
AB	322.95	S 56° 33′ 54″ E	− 177.924	+ 269.505
BC	354.85	N 34° 06′ 42″ E	+ 293.797	+ 199.003
CD	219.95	N 57° 12′ 02″ W	+ 119.147	− 184.884
DE	129.35	S 83° 34′ 04″ W	− 14.491	− 128.536
EA	269.95	S 35° 00′ 00″ W	− 221.130	− 154.837
		Total error	− 0.619	+ 0.250

6.7.5 Correction of rectangular co-ordinates

The latitudes and departures of the traverse do not close at Station A but at a point 0.619 m south of A (negative latitude error) and 0.25 m east of A (positive departure error) to give a closing error of AA′

(The closing error)^2 = $0.25^2 + 0.619^2$ (Pythagorus)

Closing error   = $\underline{0.668\,\text{m}}$

The traverse survey

This closing error of 0.668 m constitutes an error over the total horizontal distance of:

$$\frac{\text{Closing error}}{\text{Total length of survey}} = \frac{0.668}{1297.05} = \frac{1}{1941.69}$$

which is within the limits of accuracy for a site survey.

Where this error is unsatisfactory for the level of work being undertaken then the source of the error must be identified and corrected or the fieldwork repeated. Where a gross error has been identified and one line is considered incorrect the bearing of that line should also be in error and this may reduce the checking procedures if the bearing is identified.

A number of procedures are available for correcting the traverse and these are discussed in advanced surveying books. One method is Bowditch's method which was outlined graphically for the compass traverse.

Bowditch's method of correction. The method assumes that any errors in the measured distance are proportional to the square root of the measured distance and apportions the difference in the eastings and northings to each traverse leg in the ratio of the length of each leg as follows:

$$\text{Correction to the latitude} = \text{Total latitude correction} \times \frac{\text{length of leg}}{\text{total length of traverse}}$$

$$\text{Correction to the departure} = \text{total departure correction} \times \frac{\text{length of leg}}{\text{total length of traverse}}$$

The corrected co-ordinates will adjust the bearings of each station and this may not be justified since the error in linear measurement will usually be greater than the angular measurements obtained by the theodolite. This assumption was not true for the compass traverse where the graphical solution was outlined, since the angular errors were large. The graphical solution does not satisfy the levels of accuracy demanded by the theodolite traverse and the calculations shown in Table 6.4 are required.

The sum of the corrections in each case is equal to the total error and this will provide an arithmetical check of the work before correcting the latitudes and departures. To correct the latitudes and departures the latitude correction must be **added** since this is a negative error and the departure error **subtracted** as this is a positive error. This is shown in Table 6.5.

Example of a theodolite traverse calculation

Table 6.4 Bowditch's correction

Leg	Leg length perimeter of traverse	Latitude correction (error = 0.619)	Departure correction (error = 0.250)
AB	$\frac{269.95}{1297.05} = 0.248$	0.154	0.062
BC	$\frac{354.85}{1297.05} = 0.274$	0.170	0.069
CD	$\frac{219.95}{1297.05} = 0.170$	0.105	0.043
DE	$\frac{129.35}{1297.05} = 0.100$	0.062	0.024
EA	$\frac{269.95}{1297.05} = 0.208$	0.128	0.052
	Total error	0.619	0.250

Table 6.5 Corrected latitudes and departures

Leg	Correction to latitude	Corrected latitude	Correction to departure	Corrected departure
AB	+0.154	−177.788	−0.062	+269.443
BC	+0.170	+293.967	−0.069	+198.934
CD	+0.105	+119.252	−0.043	−184.927
DE	+0.062	− 14.429	−0.024	−128.560
EA	+0.128	−221.002	−0.052	−154.889
	Sum of latitudes	0.000	Sum of departures	0.000

The bearing of each leg (Table 6.5) has been adjusted, assuming that the error in co-ordinate closure was due to discrepancies in both the linear and angular measurements. Where the error can be attributed to the linear measurement, other more complex adjustment procedures are available and after the initial correction of the internal angles all subsequent adjustments leave the bearings unaltered.

The method of adjustment selected must take the following into account.

(a) the final plot should be the closest to the original results;
(b) the procedure selected must be consistent with the survey equipment used;
(c) the adjustment procedures should be consistent with the accuracy demanded by the survey application.

The traverse survey

6.7.6 Calculation of the rectangular co-ordinates

The rectangular co-ordinates reference each station to the origin and are calculated from the partial co-ordinates, once they have been corrected. Station A has been given the arbitrary co-ordinates of 100 m East and 200 m North so that all the rectangular co-ordinates will be positive. The co-ordinates of the reference station may relate to absolute locations or be chosen arbitrarily to aid the calculations.

Table 6.6 illustrates the procedure and provides one line for the **rectangular co-ordinates** of each **station** with alternate lines provided for the partial co-ordinates of the intermediate legs. The rectangular co-ordinates are obtained by **adding** (taking account of the positive and negative values) the partial co-ordinates (latitudes and departures) of each leg to the previous station's rectangular co-ordinates. Note that the **first** station is included at the end of the table to provide a check on the calculations.

Once the rectangular co-ordinates have been calculated and the last set of co-ordinates in the table are equal to the reference co-ordinates of the origin, then the traverse may be plotted.

Table 6.6

Leg	Corrected partial co-ordinates		Rectangular co-ordinates	
	Departure	Latitude	Easting	Northing
A			100.00	100.00
AB	+269.443	−177.787		
B			369.443	22.212
BC	+198.934	+293.967		
C			568.377	316.179
CD	−184.05	+119.252		
D			383.450	435.431
DE	−128.560	−14.429		
E			254.889	421.002
EA	−154.889	−221.002		
A			100.000	200.00

6.8 PLOTTING THE TRAVERSE

The co-ordinated stations are plotted from an accurately plotted grid. When setting out the grid, the lines must be parallel and at right angles to each other with the spacing accurate. To achieve this desired accuracy, the vertical lines should be set out on the top and bottom of a square, whose diagonals have been accurately checked; a similar procedure is

Methods of reducing the calculations

then followed for the horizontal grid lines. Alternatively, grid sheets are available for a wide range of popular scales with the grid intersections clearly indicated. The reference station is located on the sheet, ensuring that all the survey is capable of being contained on the sheet and the stations plotted from this reference by means of the rectangular coordinates. The accuracy of the plotting should be verified by scaling each traverse leg and checking the result with the fieldwork.

Once the control stations have been accurately located the detail can be plotted using offsets from the respective traverse lines. Where details have been referenced to more than one traverse line the scaled distances may be checked against the field notes to check the plotting accuracy.

6.9 METHODS OF REDUCING THE CALCULATIONS

The traditional procedures of using six or seven figure logarithm tables or traverse tables which list the **easting** and **northing** of a traverse leg measuring from 1 m to 10 m or 2 m to 100 m in increments of 1 minute of bearing are being replaced by:

Programmable calculators. These are modern calculators which include trigonometric function keys which enable whole circle bearings to be directly inputed for the calculation of latitudes and departures, eliminating the need to determine reduced bearings. Where a negative sign is displayed this must be ignored since it does not relate to the convention used in surveying. The results produced will be accurate to **nine** or more decimal places, which satisfies the field accuracy outlined in this section.

Programmable calculators which include trigonometrical functions often accommodate 128 program steps with up to 11 memories for storing data and these may be used to display the partial co-ordinates of each station once the following data has been input;

(a) the length of the traverse leg;
(b) the whole circle bearing.

Care must be taken when programming the above routine since the mathematical convention is to measure angles anti-clockwise from the **horizontal** while bearings are measured **clockwise** from the **vertical** axis and relate the positive or negative sign to the trigonometrical function. By including a step in the program which adds 450° to the whole circle bearing the positive or negative sign in the display will identify the direction of the partial co-ordinates.

Calculators which include a nestable loop facility for sub-routines and adequate memory capacity are capable of more advanced program applications and may accommodate the correction of co-ordinates for a limited traverse, presenting the results on hard copy.

The traverse survey

Input sheets should be devised for all the above applications which allow the field data to be collected in the format required for input at each stage of the program and the output results to be recorded, where a printer is not used.

Micro-processors. The table of co-ordinates in Table 6.7 was provided by a micro-processor program which asks the user for the following information:

(a) the included angle at each station;
(b) the length of each traverse leg;
(c) the bearing of one leg of the traverse;
(d) the reference co-ordinates of the station from which the bearing was observed;
(e) indication if the traverse is to be plotted clockwise or anti-clockwise at the start of the package and a check of the data input before any calculations are undertaken.

From the input data a print-out is provided (Table 6.7), listing the input data and the results of the following calculations:

(a) corrected angles;
(b) whole circle bearings;
(c) partial co-ordinates;
(d) corrected rectangular co-ordinates;
(e) area of the traverse.

The accuracy of the fieldwork and the relevance of the results may be estimated from the listing of:

(i) correction to angles, stated in seconds;
(ii) closing error of partial co-ordinates.

This type of facility provides a speedy solution for the surveyor since the program illustrated was run in 6 minutes 54 seconds which includes the input of the initial data. This type of package can be operated on a small 8 kb machine providing immediate results for traverse plotting.

Computer bureau facilities. Many organisations have developed specialised packages for the traverse survey, having the ability to input standard field notes and the maximum acceptable error for the specific application. From this information the rectangular co-ordinates are calculated with the ability to select a range of correcting procedures and:

(a) a plot of the survey stations;
(b) reduction of measured distances to the transverse mercator projection used by the Ordnance Survey;
(c) where reduced levels have been included, packages are available which provide contour maps, cross-sections and isometric projections if required.

Methods of reducing the calculations

Table 6.7 Traverse produced by a micro-processor package

FIELD SURVEY STATION	LENGTH (M)	ANGLE DEG:MIN:SEC	COR.ANGLE DEG:MIN:SEC	W.C.B. DEG:MIN:SEC	CO-ORDINATES EASTINGS	NORTHINGS	RECTLR CO-ORDINATES EASTINGS	NORTHINGS
A							0	0
	322.95	88 : 30 : 10	88 : 26 : 6	123 : 26 : 6	269.442835	-177.788125	269.442835	-177.788125
B		90 : 44 : 40	90 : 40 : 36	34 : 6 : 42	198.933755	293.96623	468.37659	116.178105
	354.85							
C		88 : 45 : 20	88 : 41 : 16	302 : 47 : 58	-184.926438	119.252013	283.450152	235.430118
	219.95							
D		140 : 50 : 10	140 : 46 : 6	263 : 34 : 4	-128.560838	-14.428996	154.889314	221.001121
	129.35							
E		131 : 30 : 0	131 : 25 : 56	215 : 0 : 0	-154.889314	-221.001121	0	0
	269.95							
A								

AREA OF TRAVERSE = 109045.8 SQUARE M   TOTAL LENGTH OF TRAVERSE = 1297.05 M

CORRECTION TO ANGLES = 244 SECONDS

ERROR IN TRAVERSE = 1 IN 1939
PROGRAM RUN IN 6 MINUTES; 54 SECONDS

Table 6.8 Traverse table

Traverse table — Title — Date

Station	Field and corrected angles	Leg length (m)	WCB	RB	Calculations	Partial co-ordinates Uncorrected				Partial co-ordinates Corrected				Rectangular co-ordinates	
						Departure		Latitude		Departure		Latitude			
						E	W	N	S	E	W	N	S	E	N
A	88° 30′ 10″ 88° 26′ 06″	322.95	123° 26′ 06″	S 56° 33′ 54″ E	322.95 sin 56° 33′ 54″ 322.95 cos 56° 33′ 54″	269.505			177.942	269.443			177.788	100.000	200.000
B	90° 44′ 40″ 90° 40′ 36″	354.85	34° 06′ 42″	N 34° 06′ 42″ E	354.85 sin 34° 06′ 42″ 354.85 cos 34° 06′ 42″	199.003		293.797		198.934		293.967		369.443	22.212
C	88° 45′ 20″ 88° 41′ 16″	219.95	302° 47′ 58″	N 57° 12′ 02″ W	219.95 sin 57° 12′ 02″ 219.95 cos 57° 12′ 02″		184.884	119.147			184.927	119.252		568.377	316.179
D	140° 50′ 10″ 140° 46′ 06″	129.35	263° 34′ 04″	S 83° 34′ 04″ W	129.35 sin 83° 34′ 04″ 129.35 cos 83° 34′ 04″		128.536		14.491		128.560		14.429	383.450	435.431
E	131° 30′ 00″ 131° 25′ 56″	269.95	215° 00′ 00″	S 35° 00′ 00″ W	269.95 sin 35° 00′ 00″ 269.95 cos 35° 00′ 00″		154.837		221.130		154.889		221.002	254.889	421.002
A	88° 30′ 10″ 88° 26′ 06″													100.000	200.000

6.10 THE TRAVERSE TABLE

An example of a traverse table is outlined in Table 6.8. Whilst other formats are available, the following requirements are illustrated:

(a) Column 2 has provision for the insertion of the mean angle observed at each station and the corrected angle. After the mean angles have been corrected the new value may be inserted in red or the mean angle cancelled out.
(b) Column 3 is staggered between the lines for each station since this refers to the intermediate legs.
(c) Column 4 may be completed after the WCBs have been calculated.
(d) The data for columns 7–10 may be calculated directly and inserted in the respective columns. Since the partial co-ordinates relate to the traverse legs these are on the same line as the leg lengths.
(e) After applying the respective corrections to the partial co-ordinates, the rectangular co-ordinates may be determined and entered in columns 15 and 16. Note that the entries in columns 15 and 16 are in line with the respective stations in column 1.

Example 6.4

The details of a short theodolite survey between two points A and F are given in Table 6.9. The survey is to be plotted by the use of rectangular co-ordinates. Calculate the co-ordinates and bearing of AF. Determine the distance of AF. (RICS)

Table 6.9

Line	Bearing	Distance
AB	N 22° 03' E	80.90
BC	S 83° 37' E	63.80
CD	N 19° 48' E	66.20
DE	N 61° 12' E	70.10
EF	N 3° 06' W	146.70

1. The data provide the reduced bearings which allow the latitudes and departures of each station to be calculated.
2. From the latitudes and departures of each traverse leg the rectangular co-ordinates of each station can be determined. Assume the reference co-ordinates of Station A to be 0 m East, 0 m North, since this will assist the calculation of distance AF and no other values have been given.

The bearing of AF from Station A can be obtained from the co-ordinates of Station F relative to Station A, since these co-ordinates form a right-angled triangle.

The traverse survey

Table 6.10

Station	Line	Length	Bearing	Latitude	Departure	Co-ordinates	
						Northing	Easting
A						0	0
	AB	80.90	N 22° 03′ E	+ 74.98	+ 30.37		
B						74.98	30.37
	BC	63.80	S 83° 37′ E	− 7.09	+ 63.40		
C						67.89	93.77
	CD	66.20	N 19° 48′ E	+ 62.29	+ 22.42		
D						130.18	116.19
	DE	70.10	N 61° 12′ E	+ 33.77	+ 61.43		
E						163.95	177.62
	EF	146.70	N 3° 06′ W	+ 146.48	− 7.93		
F						310.43	169.69
	FA						
A						0	0

Tangent of bearing of AF = departure/latitude.

$$\text{Bearing of AF} = \tan^{-1} \frac{169.69}{310.43}$$

$$= \tan^{-1} 0 - 546\,63$$

$$= 28° 39′ 44″$$

Bearing = N 28° 39′ 44″ E.

The length of leg AF may be determined from the same right-angled triangle, since

$$AF^2 = \text{easting}^2 + \text{northing}^2$$
$$AF = 169.69^2 + 310.43^2$$
$$= 353.78\,m$$

6.11 EXERCISES

1. Two points A and E, which are not intervisible, are on the centre line of a straight roadway which is to be constructed.

   The details of a traverse made between the points A and E are as follows:

Line	Azimuth or bearing	Distance (metres)
AB	89° 00′	597.00
BC	170° 00′	390.00
CD	181° 00′	550.00
DE	280° 00′	355.00

The survey is to be plotted by the use of rectangular co-ordinates. Calculate the rectangular co-ordinates. So that the cost of making the road-way may be known, calculate the distance AE and the bearing of that line. (RICS)

2. The following measurements refer to a closed traverse, A B C D E A. Taking the co-ordinates of A as (0, 0) and the whole circle bearing of AB as 000° 00′ 00″ calculate the corrected co-ordinates of the traverse stations.

Clockwise angle	Measured value
EAB	270° 00′ 03″
ABC	235° 20′ 38″
BCD	279° 42′ 43″
CDE	239° 06′ 23″
DEA	235° 50′ 28″

Line	Horizontal length in metres
AB	38.971
BC	121.687
CD	26.063
DE	102.108
EA	53.784

(RICS)

3. Describe in detail the method of carrying out a traverse by the use of a theodolite and steel tape. State a method to be used for measuring the horizontal angles and why this has been adopted. Give an example of booking the horizontal angles. (RICS)

7 THE LEVEL

7.1 THE PRINCIPLES OF LEVELLING

The procedure of levelling adopted for most survey applications will require a horizontal line to be formed against which vertical measurements can be made, in order to determine the **relative** heights of other locations. The principle is illustrated in Fig. 7.1 where a horizontal line is 101.25 m above the selected datum. The respective vertical distance between this reference line and each station is then subtracted from the height of the horizontal line to obtain the **reduced** level of each station (see section 8.1).

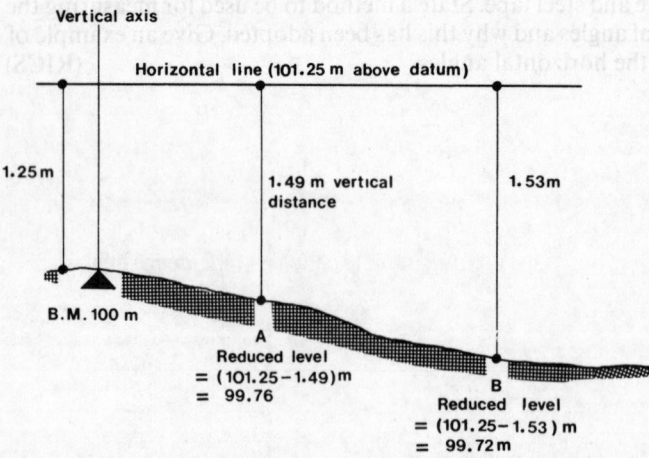

Fig. 7.1 Principles of levelling

7.2 COMPONENTS OF THE LEVEL

The level is composed of the following elements.

7.2.1 The bubble tube

Gravitational levelling is the procedure used by surveyors where the horizontal reference line (Fig. 7.1) is defined by some form of spirit level arrangement. The simplest spirit level will only prove accurate where the level is to be transferred over a distance of approximately 1 m, therefore the range must be increased by means of a telescope.

The bubble tube (Fig. 7.2) is constructed of glass and is circular in cross-section with the longitudinal section forming part of a circular arc. The tube is filled with a low viscosity liquid, eg alcohol or sulphuric acid containing an air bubble. When the bubble floats in the centre of the tube, the tangent of the tube is horizontal and the reference line defined.

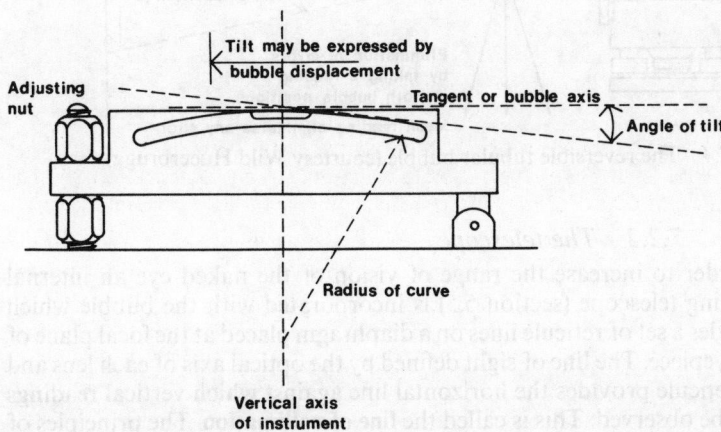

Fig. 7.2 The bubble tube

The sensitivity of the bubble may be increased by placing a prism arrangement at each end of the bubble tube, as illustrated in Fig. 7.3. This enables both ends of the bubble to be seen simultaneously, and as the instrument is tilted the two halves of the bubble appear to move in opposite directions. This arrangement is called a **coincidence bubble** since

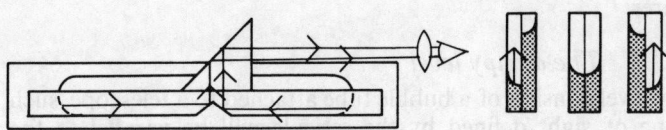

Fig. 7.3 The coincidence bubble

The level

centring is achieved by bringing the two ends of the bubble into coincidence (two halves opposite each other) when viewed.

The bubble accuracy may also be increased by using a barrel-shaped tubular bubble having the same curvature both at the top and the bottom. This allows the telescope to be rotated about its horizontal axis, enabling the bubble to be reversed. This arrangement (Fig. 7.4) allows two readings to be observed at each staff position. The mean of the two readings will be of a high accuracy due to the compensation for the instrument errors outlined in section 7.8.

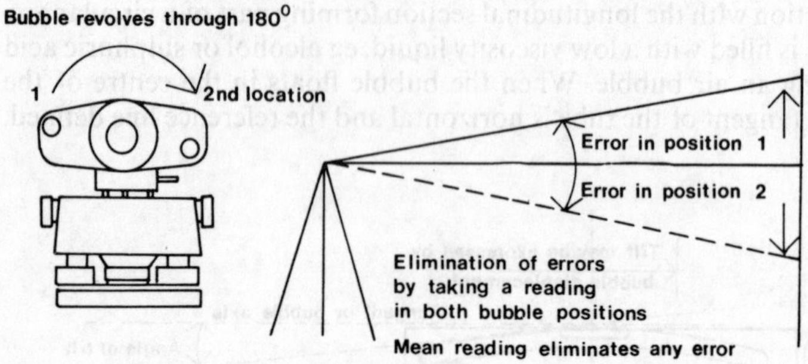

Fig. 7.4 The reversible tubular bubble (courtesy Wild Hueerbrugg)

7.2.2 The telescope

In order to increase the range of vision of the naked eye an internal focusing telescope (section 5.5) is incorporated with the bubble which includes a set of reticule lines on a diaphragm placed at the focal plane of the eyepiece. The line of sight defined by the optical axis of each lens and the reticule provides the horizontal line against which vertical readings may be observed. This is called the **line of collimation**. The principles of the internal focusing telescope are outlined in section 5.5 and the adjustment procedures for eliminating parallax are exactly the same as for the theodolite.

7.3 TYPES OF LEVEL

There are three basic arrangements of the telescope and bubble, although slight variations occur within each group depending on the manufacturer.

7.3.1 The dumpy level

The dumpy level consists of a bubble tube attached to a telescope, such that the line of sight defined by the reticule will be **parallel** to the bubble axis. The telescope is mounted on a tribrach incorporating three

Types of level

foot screws, which enable the instrument to be levelled. Once the bubble has been brought to the centre of its axis, the bubble axis and the line of collimation will be horizontal and at right angles to the vertical axis (Fig. 7.5). When the instrument is rotated to any horizontal position about the vertical axis any number of staff readings can then be obtained without further adjustment.

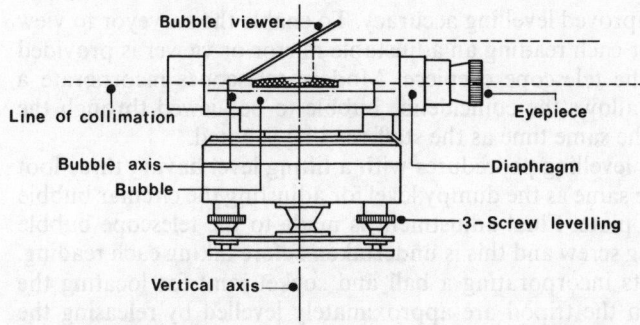

Fig. 7.5 The dumpy level

When setting up the instrument or undertaking **temporary adjustment** – which is the term used to describe the procedure of ensuring that the vertical axis is located over the selected instrument position and the bubble is horizontal – the following procedures are followed:

(a) After locating the tripod over the station using the procedures outlined in section 5.8.2 for the theodolite, screw the tribrach of the level on to the tripod head.
(b) If a circular bubble is located on the horizontal plate of the tribrach this may be used to roughly level the lower plate while centring the instrument over the station. This is undertaken by adjustment of the tripod legs as outlined for the theodolite (section 5.8.2) and observing the plumb bob attached to the underside of the instrument that remains over the station.
(c) Final adjustment of the instrument is achieved by rotating the telescope over two foot screws and bringing the bubble to the centre of its run by turning **both** the foot screws inwards **or** outwards. The bubble will be found to move in the direction of the **left** thumb.
(d) Turn the telescope through 90° so that it lies over the third foot screw, which is then used to set the bubble in the centre of its run. Only move this foot screw when making this adjustment.
(e) Repeat procedures (c) and (d) until the bubble is central for all positions of the instrument allowing observations to be recorded in any horizontal direction.

The level

7.3.2 The tilting level

The telescope and bubble are located on the tripod by means of a ball and socket joint or a three-screw levelling system. The telescope is capable of being tilted in the vertical plane by means of a **tilting screw** at the eyepiece. This tilting screw is used to bring the telescope bubble horizontal **each time** a reading is observed (Fig. 7.6).

Since the instrument is adjusted for each reading it may be used for the highest levels of accuracy and usually incorporates a coincidence bubble for improved levelling accuracy. To enable the surveyor to view the bubble for each reading an adjustable mirror or viewer is provided adjacent to the telescope eyepiece. Modern telescopes incorporate a prism which allows the coincidence bubble to be viewed through the telescope at the same time as the staff reading is noted.

The initial levelling procedures with a tilting level having three foot screws are the same as the dumpy level for adjusting the circular bubble on the upper plate. Final adjustment is made to the telescope bubble with the tilting screw and this is undertaken **before** taking each reading.

Instruments incorporating a ball and socket joint for locating the instrument on the tripod are approximately levelled by releasing the joint and carefully rotating the instrument until the circular bubble is central. Final adjustment to the telescope bubble is again undertaken by means of the tilting screw for each observation.

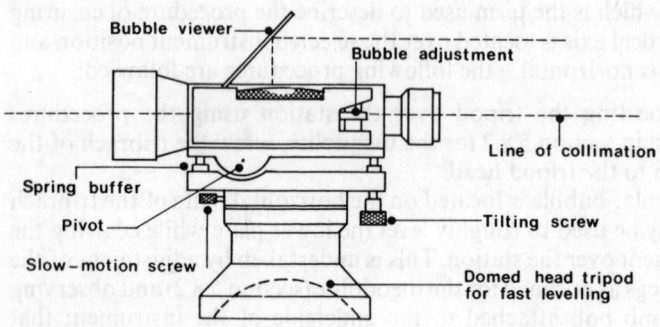

Fig. 7.6 The tilting level

7.3.3 The automatic level

This type of level incorporates a self-levelling compensator located inside the optical system of the telescope. The **compensator** consists of a pendulum and prism arrangement (see Fig. 7.7) which automatically sets the line of collimation horizontal within ± 15 minutes of tilt. The tribrach may consist of a three-screw levelling device or a ball and socket arrangement which must be levelled within the range of the compensator, using the procedures described previously for each tribrach.

Accuracy of levels

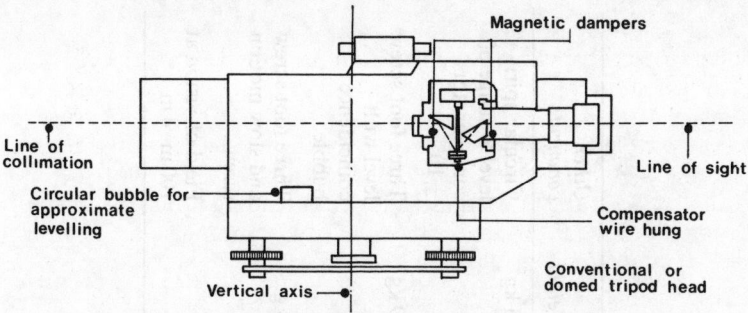

Fig. 7.7 The automatic level (courtesy Carl Zeiss Zena Ltd)

By including a compensator arrangement within the telescope an erect image will be presented and air or magnetic dampers are often incorporated to produce a stable image and prevent prolonged oscillation of the compensator when taking a reading.

Initial levelling using either the three-screw or ball and socket arrangement is monitored on a circular bubble incorporated on the upper plate. The instrument does not include a telescope bubble since the line of collimation is automatically set once the compensator is operating. To ensure that the compensator is operating correctly the instrument should be gently tapped and the image observed to check that it oscillates before becoming stable. Many instruments incorporate coloured filters which prevent an observation being made if the compensator is not operating.

7.4 ACCURACY OF LEVELS

This is normally related to one of the following:

Class 1. Geodetic or similar forms of precise levelling require a Class 1 instrument with a very sensitive bubble having a telescope of high magnification and a large aperture measuring to ± 0.2 secs of arc. The instrument is usually a tilting or some form of automatic level.

Class 2. These instruments are used for medium accuracy work and have a magnification of $20 \times - 40 \times$ with an aperture of 30 mm–40 mm being capable of levelling to 0.5–1 second of arc.

Class 3. This group includes the builder's level, which is used for comparatively low accuracy work of short range using an instrument which provides $10 \times - 20 \times$ magnification having a lens diameter of 20 mm to 30 mm. The instrument may be a dumpy or tilting level.

A summary of the accuracy of a range of instrument can be found in Table 7.1.

Table 7.1 Instrument accuracy

Instrument	Magnification	Sighting distance	Horizontal circle	Accuracy	Weight	Other comments
Automatic precise level	31.5×	2.2 m to 140 m	360° reading to 10′ estimating to 1′	±0.5 mm to ±5 mm at 350 m. Compensator ±0.15″ arc	4.1 kg	Circular spirit level 8′ range of compensator ±10′
Engineer's level	25×	2 m to 100 m	360° reading to 10′ estimation to 1′	1 mm up to 1 cm at 350 m	1.9 kg	Three foot screw level with coincidence bubble
Engineer's automatic level	25×	1.5 m to 100 m	As above	1 mm up to 0.5 cm at 350 m	1 kg	Three foot screw and slow motion screw
Quickset level	20× 30 mm aperture	0.75 m to 100 m	360° reading to 20 minutes	1 mm up to 1 cm at 250 m		Field of vision at 100 m–4 m

7.5 THE SPECIFICATION OF THE LEVEL

The levels outlined indicate a wide range of instruments available and the surveyor's choice will depend on the accuracy of the work being undertaken. The accuracy of the instrument will be influenced by:

The telescope power. The power of the telescope will depend on the magnification of the object and the size of the object lens (aperture).

The support system. These may be of the ball and socket type where accuracy is provided by adjusting the bubble for each reading. This form of support is used for the more accurate levels and where a series of readings are observed from different instrument positions. The three-foot screw levelling device may take longer to set up but it will prove satisfactory for most builder's work where a number of readings are observed from one instrument position. Instruments requiring a high level of accuracy combine both of these support systems.

Refinement of control. These should be selected with regard to the accuracy required and may include:

(a) horizontal circle adjustable to zero;
(b) coincidence bubble for levelling;
(c) slow motion screws for sighting onto the object;
(d) erect image telescope;
(e) horizontal circle reading by means of a magnifier;
(f) colour filters when the instrument is operating outside the normal levels of illumination;
(g) a tribrach capable of accommodating other instruments.

7.6 USE OF THE LEVEL

All levels are free to rotate about the **vertical axis** of the instrument, and after initial levelling and accurate location over the instrument station the following procedure is adopted:

1. The telescope is rotated on to the staff by sighting along the top of the telescope barrel to the staff position.
2. Lock the horizontal motion of the instrument and focus the telescope on to the staff by means of the focusing screw. **Do not** move the **eyepiece** since this is already focused on to the diaphragm to eliminate **parallax** (section 5.5.4).
3. Using the slow motion screw to control the horizontal movement of the instrument, **bisect** the staff with the intersection of the vertical and central horizontal reticule. The **vertical** reticule may be used to check that the staff is being held correctly.

The level

4. Check that the instrument bubble is **central** and note the staff reading at the intersection of the **centre** stadia mark.
5. Book the reading and **check** the bubble position and staff reading at least twice.
6. Turn the instrument to sight on to the next staff position and repeat operations 1–6.

After observing all the staff readings at one instrument position, the staff will **remain** at the same location and the instrument is moved as follows:

(a) check that the instrument is firmly located on the tripod;
(b) close the bubble mirror (if fitted) and place the lens cover firmly over the telescope;
(c) close the tripod legs and, firmly holding the legs together, carry the instrument to the next instrument station along with the instrument box, containing the plumb bob, etc;
(d) **never move the instrument and staff together.**

On completion of the fieldwork the instrument should be carefully cleaned and dried with a soft cloth. The following should be observed before replacing the level in its case:

(a) centre the foot screws (if fitted);
(b) close the bubble mirror cover and place the lens cover over the telescope;
(c) release the horizontal clamp and remove the level from the tripod;
(d) place the instrument carefully in its box, ensuring that the telescope and tribrach are correctly located. Close the cover and check that the cover and securing strap is firm;
(e) place the cover over the head of the tripod and clean the legs with a soft cloth;
(f) close the tripod legs and strap them together, ensuring that the locking screws on each leg are tight.

7.7 THE LEVELLING STAFF

This provides the means of measuring the vertical distance from the line of collimation to the ground level at the location under consideration. The staff is usually 4 m long, made of aluminium alloy and may be telescopic, with each section slotting together or folding for transit. Care must be taken when using both of these arrangements to ensure that they have been correctly extended or unfolded when a reading is observed.

The staff may have a white or yellow epoxy resin face, graduated in 10 mm intervals with the odd numbered graduations in red and the even numbers in black. BS 48484: Part 1: 1969 recommends that the lower 50 mm of each 100 mm section be of a distinct E formation to aid reading

The levelling staff

when viewed through the telescope. Many modern staffs have the graduations and figures inverted, to provide an erect image when the staff is viewed through the telescope as illustrated in Fig. 7.8.

Care must be taken when observing a staff reading that the staff is vertical, and whilst this can be determined by the vertical reticule of the level for one direction, spot bubbles are available for locating on the staff to assist the staff man control verticality in the other direction.

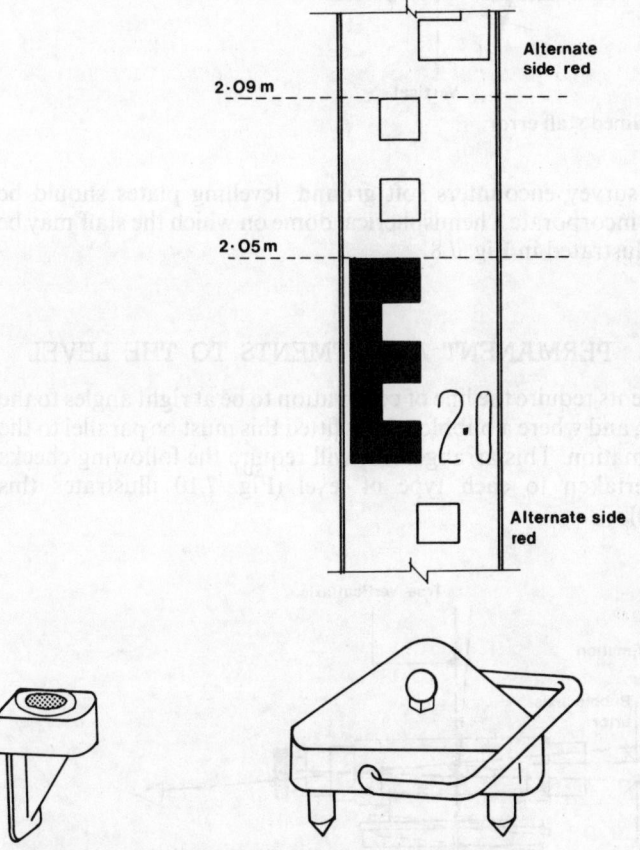

Fig. 7.8 The levelling staff, bubble and levelling plate.

Example 7.1 Inclined staff error

Determine the correction to be made when a staff is held 5° off verticle

$$\text{True staff reading} = (\text{observed staff reading}) \times \cos 5°$$
$$\text{Correction} = -(1 - \cos 5°) \times \text{staff reading}$$
$$= \underline{0.0038 \times \text{staff reading}} \text{ (see Fig. 7.9)}$$

The level

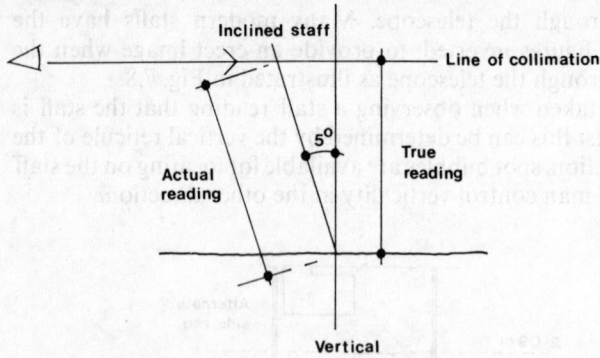

Fig. 7.9 Inclined staff error

Where the survey encounters soft ground, levelling plates should be used. These incorporate a hemispherical dome on which the staff may be placed as illustrated in Fig. 7.8.

7.8 PERMANENT ADJUSTMENTS TO THE LEVEL

All adjustments require the line of collimation to be at right angles to the vertical axis, and where a bubble tube is fitted this must be parallel to the line of collimation. This arrangement will require the following checks to be undertaken to each type of level (Fig. 7.10 illustrates this relationship).

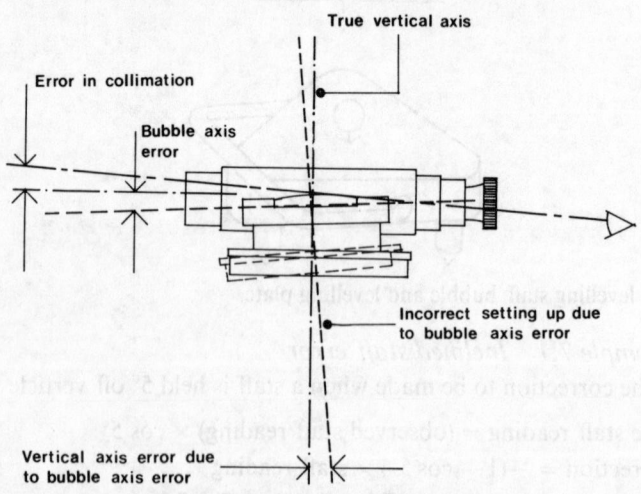

Fig. 7.10 Permanent adjustments to the level

7.8.1 The dumpy level

The bubble axis must first be set perpendicular to the vertical axis, since this is used to set up the instrument. The following procedures are therefore undertaken:

(a) level up the instrument as accurately as possible;
(b) turn the telescope so that it lies parallel to two of the foot screws;
(c) bring the bubble to the exact centre of the bubble tube using the foot screws;
(d) turn the telescope through 180° horizontally and observe if the bubble remains central;
(e) if the bubble moves off centre bring it **half** way back with the foot screws and the other half by the bubble adjusting screw under the bubble tube (Fig. 7.2);
(f) turn the telescope through 90° and centre the bubble with the **third** foot screw.

Once the bubble has been set perpendicular to the vertical axis the line of sight may be set parallel to the bubble axis **and** at right angles to the vertical axis as follows:

(a) set up the instrument **mid-way** between two pegs driven into the ground 60 m–80 m apart and in line when observed on plan;
(b) hold the staff on peg A (see Fig. 7.11) and note the reading (say, 1.400 m);
(c) hold the staff on peg B until the same reading is observed on both pegs A and B (say, 1.40 m);
(d) move the instrument to instrument position 2 which is close to one of the pegs (see Fig. 7.11) and take a reading on to the staff on peg A;
(e) when a reading is taken through on to the peg at Station B, the staff reading should be the same as reading (d) on peg A;

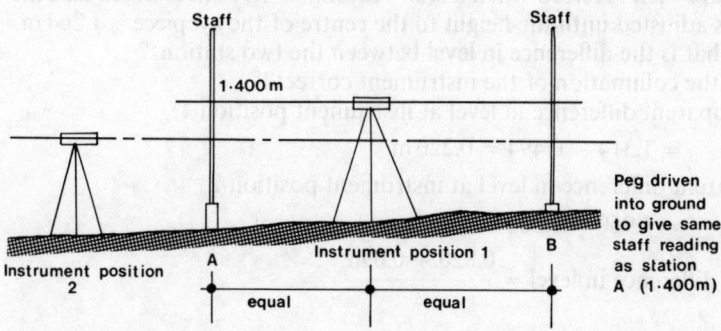

Fig. 7.11 The two peg test

The level

(f) if there is a difference between the two readings then the upper and lower diaphragm screws on the telescope must be adjusted until the horizontal reticule gives the same reading at stages (d) and (e) of the procedure.

Note that the pegs will be correct at stages (b) and (c) since the instrument is equidistant between the two stations, which means that any error will be of the same magnitude in both directions and the pegs will be **level** (see Example 7.2).

The above test, which is carried out **after** the bubble axis has been adjusted, is often referred to as the **two peg test** and enables the line of collimation to be set parallel to the bubble axis of the instrument.

7.8.2 The tilting level

The tilting level has only one permanent adjustment since it is only necessary to check that the line of collimation and bubble axis are **parallel** because the bubble axis is adjusted for each reading. This can be verified by undertaking the **two peg test** as outlined for the dumpy level. In the case of the tilting level, the bubble tube is adjusted rather than the diaphragm after the telescope has been sighted on to the correct staff reading by means of the tilting screw.

7.8.3 The automatic level

The only check to be undertaken with the automatic level is the **two peg test** to ensure that the line of collimation is horizontal when the instrument has been levelled within the limits of the compensator. Any corrections are normally carried out by adjustment of the **diaphragm**. Any other adjustment that may be required must be undertaken by the manufacturer.

Example 7.2

A dumpy level is over a station and the height from ground level to the centre of the eyepiece is found to be 1.314 m. A level is then held on a distant peg and a staff reading of 0.494 m is noted. The instrument and staff are then reversed when a staff reading of 1.960 m is noted and the peg is adjusted until the height to the centre of the eyepiece is 1.264 m.

What is the difference in level between the two stations?
Is the collimation of the instrument correct?
Apparent difference in level at instrument position 1

$$= 1.314 - 0.494 = 0.220 \, \text{m}$$

Apparent difference in level at instrument position 2

$$= 1.960 - 1.264 = 0.696 \, \text{m}$$

$$\text{True difference in level} = \frac{0.820 + 0.696}{2}$$

$$= 0.758 \, \text{m}$$

Since the apparent difference in level is not equal to to the true difference in level the collimation of the instrument is not correct. When the level is located over the peg, the staff reading will be

$$x - 1.264 = 0.758 \text{ m}$$
$$x = \underline{2.022 \text{ m}}$$

The instrument may be corrected by adjusting the diaphragm to give this staff reading.

Alternatively, the distance between the staff readings may be obtained, say 25 m, and all further readings corrected by a value of $(2.022 - 1.96)/50 = \underline{0.062 \text{ m in } 50 \text{ m}}$

Example 7.3

A level is set up 25 m from peg A and 50 m from peg B. The respective readings on staffs A and B were 1.714 m and 2.037 m. It was known that the reduced level of peg A was 35.375 m and 35.080 m respectively. Find:

(a) the collimation error;
(b) the true readings on each staff.

The staff readings indicate that staff B is lower than staff A, by:

$$2.037 - 1.714 = 0.323 \text{ m}$$

The reduced levels of each staff position indicate that B is lower than staff position A, by:

$$35.375 - 35.080 = 0.295 \text{ m}$$

Therefore the collimation error = $0.323 - 0.295 = 0.028 \text{ m}$

ie 28 mm in 25 m
or 1 mm in 0.89 m.

True reading on staff B = 2.037 − 0.056 = 1.981 m.
True reading on staff A = 1.714 − 0.028 = 1.686 m.
Check: difference in staff readings = 1.981 − 1.686 = 0.295 m.
This is equal to the difference in reduced levels, therefore all the instrument readings can be corrected providing the distance to the staff position from the instrument is known.

7.9 EXERCISES

1. The permanent adjustment of a level must be checked from time to time. Describe how this is done for:

   (a) a dumpy level;
   (b) a tilting level.

The level

Use the following three-peg test results to illustrate your answer and derive any formulae used. A B C D are four stations in a straight line such that:

$$AB = BC = CD = 15\,\text{m}$$

With the instrument set at B the reading on a vertical staff at A is 1.104 m, and on a vertical staff at C it is 1.646 m. The instrument is then moved to D and readings to staffs at A and C are now 1.236 m and 1.752 m respectively. (RICS)

2. What instructions would you give to an inexperienced assistant who had to act as staff holder and chain man for a levelling survey?
3. Four points, A, B, C and D, are in a straight line such that AB is 80 m, BC is 180 m and CD is 120 m. Using a level set up at B and C observations were made on to staves at C and D and the following readings recorded:

Level position	Staff position	Staff reading
B	A	3.642
	D	1.066
C	A	3.011
	D	0.489

What is the collimation error in the level and what is the true height difference between A and D? What should the surveyor do if his level has a collimation error? (RICS)

8 LEVELLING

8.1 INTRODUCTION

Levelling is the term given to the activities undertaken when determining the height of a point on the earth's surface. These procedures are often required to identify spot heights of point locations, locate contours on maps or provide information for construction work where a level or inclined surface is required, eg excavations, drainage, etc.

The height of a point can only be determined relative to some other reference point or **datum**. The datum may be the Ordnance datum used by the Ordnance Survey which is the mean sea level at Newlyn, Cornwall, or an arbitrary datum selected for small works, eg excavations, drain laying, etc where only the relative differences in level are required (see setting out, section 12.2).

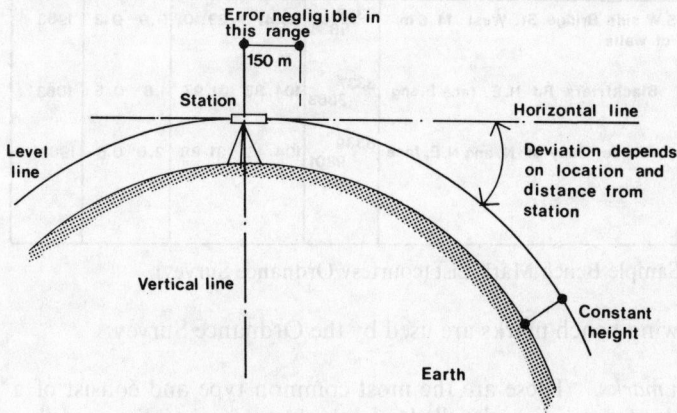

Fig. 8.1   A level and horizontal line

Levelling

A **reduced level** is the height of the point under consideration relative to the datum selected. Since the earth is curved and a level surface is one which is at a constant height above or below the datum, this must also be curved, as indicated in Fig. 8.1.

Figure 8.1 indicates the deviation between a **horizontal line** which is tangential to the level line at the station under consideration, and a **level line** which is at a constant height relative to some datum on the earth's surface and must, therefore, be a curved line since it follows the mean surface of the earth.

The deviation between these two lines is minimal for most levelling operations and may be ignored for sights under 150 m. Over this distance the corrections outlined in section 8.8(c) must be applied.

8.2 BENCH MARKS

These are reference points of known height, above a selected datum, from which the relative height of other points may be determined. Two classes of bench mark are available depending on the application of the survey:

8.2.1 Ordnance bench marks

These are permanent bench marks established by the Ordnance Survey at a known height above the Ordnance datum and are indicated on the 1/1250, 1/10 000 and 1/2500 maps or on the bench mark list supplied by the Ordnance Survey, to those who require the latest values of bench marks within a 1 km square (Fig. 8.2).

Description of bench mark	National Grid ten m. ref.	Altitude Feet	Altitude Metres	Ht.above gr Feet	Ht.above gr Metres	Date levelled
Albert Br. S.W. side Bridge St. West 14.6 m N.W. Junc. of walls	8329 9836	95.47	29.10	0.9	0.3	1963
Wks No 51 Blackfriars Rd N.E. face E.ang.	8337 9883	104.88	31.97	1.8	0.5	1963
Wall Hospl S. side Quay St N. ang N.E. face	8339 9801	104.43	31.88	2.0	0.6	1963

Fig. 8.2 Sample Bench Mark List (courtesy Ordnance Survey)

The following bench marks are used by the Ordnance Survey:

Cut bench marks. These are the most common type and consist of a horizontal reference line chiselled into the face of a building or other permanent structure with a broad arrow immediately below.

Bolt bench marks. These consist of a mushroom-headed brass bolt engraved with an arrow and the symbol 'OSBM' set in horizontal surfaces.

Flush brackets. These are metal plates fixed in the face of a building at $1\frac{1}{2}$ km intervals and have been located by geodetic levelling. Each bracket contains a serial number and a horizontal ledge from which the altitude is referenced. A detachable bracket is fixed on the plate to support a staff.

Rivet bench marks. Small brass rivets are often used on horizontal surfaces with an arrow symbol alongside and a reference symbol.

8.2.2 Temporary bench marks

These are bench marks established by the surveyor for a specific application. They are placed close to the area to be surveyed, where they will not be disturbed by the activities being undertaken. These bench marks may take an arbitrary level or can be referenced to the Ordnance datum.

8.3 LEVELLING PROCEDURES

The basic principles were outlined in section 7.1 and are illustrated in Fig. 8.3. The instrument has been set up at Station 1 so that staff readings can be observed on Stations A, B and C. The reduced levels of

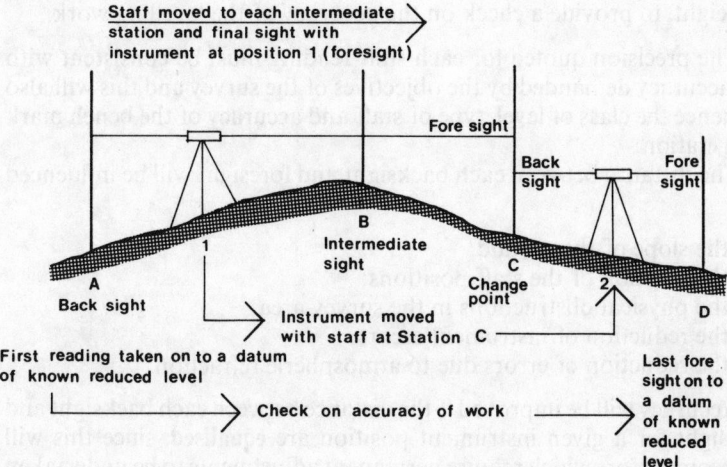

Fig. 8.3 Levelling procedures

Levelling

each staff position are determined in the following manner:

1. Station A is the first station and this is called a **backsight** since it is a point of known elevation from which the height of the line of collimation can be determined, once the staff reading has been observed. This station could relate to an arbitrary bench mark reference, depending on the purpose of the survey but all levelling procedure should commence and finish on a station of known level.
2. Note the staff readings on the remaining stations that can be observed from instrument position 1, along with any reference data, remarks, etc. Where a number of readings are observed from one instrument position these are called **intermediate sights** and are staff readings between the backsight and foresight.
3. After the staff reading at Station C has been noted the instrument will need moving to instrument position 2. The last reading taken from a particular instrument position is called a **foresight**. After a foresight reading the instrument is moved or the levelling procedures are completed.
4. When the instrument is moved to position 2 the staff remains at Station C since this will be the next backsight reading when levelling commences, being a point of known reduced level. Stations which provide both backsight and foresight readings are called **change points**.
5. After the new backsight reading has been recorded the new height of collimation can be determined and the procedures repeated. Note that when these procedures are being undertaken **either** the staff **or** the instrument may be moved but **never** both together.
6. The last staff reading of the survey will be a **foresight** and this should be taken on either the first datum or another bench mark of known height, to provide a check on the accuracy of the levelling work.

The precision quoted for each staff reading must be consistent with the accuracy demanded by the objectives of the survey and this will also influence the class of level, type of staff and accuracy of the bench mark information.

The distance between each backsight and foresight will be influenced by:

(a) the slope of the ground;
(b) the location of the staff positions;
(c) the physical obstructions in the survey area;
(d) the reduction of instrument errors;
(e) the reduction of errors due to atmospheric refraction.

Accuracy will be improved if the distance between each backsight and foresight for a given instrument position are **equalised**, since this will overcome errors which require permanent adjustments to be undertaken to the instrument (section 7.8.1).

8.4 BOOKING THE LEVELS

The levels are booked on a special booking sheet which includes three columns for the following types of reading:

1. Backsights.
2. Intermediate sights.
3. Foresights.

Each staff reading is systematically booked under the respective heading as the work proceeds with **one line for each staff position**. Any relevant comments relating to a staff station are included in the final remarks column as illustrated in Fig. 8.4.

Back sight	Inter- sight	Fore- sight	Remarks
1.155			Bench mark 20.530 A.O.D.
	1.360		Station A
	0.920		Station B
1.255		1.065	(Change pt.) C
	0.980		Station D
		0.885	Station E

Fig. 8.4 Booking procedure

Note the following:

(a) the first reading is a **backsight**;
(b) the last reading is a **foresight**;
(c) at a **changepoint** there will be a backsight and foresight;
(d) the booking procedure follows the arrows indicated for each instrument position;
(e) each station is referenced in the remarks column to relate the levels to locations on the final plan and any other relevant data may be included;
(f) a change point includes **both** a **backsight** and **foresight** since these are observed at the same staff position (the instrument **only** has been moved).

To reduce the levels one of the following procedures may be adopted:

8.4.1 The rise and fall method

The readings illustrated in the last example are entered on the booking sheet with three further columns for the determination of:

(i) the amount that a station is above the previous station (**rise**);
(ii) the amount that a station is below the previous station (**fall**);
(iii) the reduced level of each station.

Levelling

The first reading is a backsight of 1.155 m on to a bench mark of known reduced level, therefore the reduced level of this station can be entered directly as 20.530 m.

Back sight	Inter- sight	Fore- sight	Height of collimation		Reduced level	Distance (m)	Remarks
			Rise	Fall			
1.155					19.53		B.M. 19.53 A.O.D.
	1.360			0.205	19.325		Station A
	0.920		0.44		19.765		Station B
1.255		1.065		0.145	19.620		Station C
	0.980		0.275		19.895		Station D
		0 885		0.095	19.990		B.M. 19.99 m

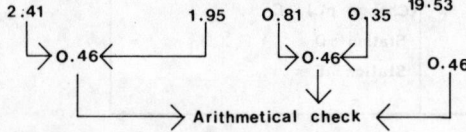

2.41   1.95   0.81   0.35   19.53

0.46   0.46   0.46

→ Arithmetical check ←

Fig. 8.5 The rise and fall method

The staff reading at Station A was recorded as 1.360 m and since this is **greater** than the previous reading of 1.155 m Station A is **lower** than the bench mark by 0.205 m and this value is entered in the **fall** column.

The next staff reading is 0.920 m at Station B, and since this is **less** than the previous reading of 1.360 m Station B must be higher than Station A by 0.44 m and this is entered in the **rise** column.

The relationship between the remaining successive stations is then determined and, when the staff reading is less than the preceding staff reading, there will be a **rise**, while a staff reading greater than the preceding reading will identify a **fall**.

At a **change point** the rise or fall of the change point is evaluated from a comparison of the foresight at that station and the last intermediate sight. The backsight at the change point is used to determine the rise or fall of the next reading after the backsight at the new instrument position. At Station D the staff reading is 0.275 m above the backsight on to Station C and this is entered in the **rise** column.

Once the rise or fall of each station relative to the previous station has been determined, the reduced levels can be evaluated. These are obtained by adding the rise of a station to the reduced level of the last station or subtracting the fall from the previous reduced level.

Booking the levels

Since the difference between the last and first reduced levels will be equal to the difference between the sum of all the rises and all the falls the arithmetical steps can be checked. The total rise or fall between each instrument position is equal to the difference between each backsight and foresight reading and this will provide a further check since the total difference will equal the difference between the first and last reduced level. The complete arithmetical check for the rise and fall procedure is:

Difference between   Difference between   Difference between
last and first   = sum of rises and   = sum of backsights
reduced level   sum of falls   and sum of foresights

This will provide a complete check of the arithmetic and should be undertaken for each page of the survey in the field. Where errors are detected these should be corrected by putting a line through the error. This check will not identify incorrect readings, therefore the first and last reading should relate to a station of known height to identify any booking errors.

8.4.2 Height of collimation method

This method relates to the levelling principles outlined in section 7.1 where the height of a station below the line of collimation is determined.

The height of collimation for each instrument position is determined from the backsight reading and the remaining reduced levels evaluated from this by subtracting the staff reading. At a change point, the last foresight will allow the reduced level at that station to be determined and the backsight on to the same staff position will allow the new height of

Back sight	Inter- sight	Fore- sight	Rise / Fall Height of collimation	Reduced level	Distance (m)	Remarks
1.155			21.685	20.530		B.M. 20.530 A.O.D.
	1.360			20.325		Station A
	0.920			20.765		Station B
1.255		1.065	21.875	20.620		Station C
	0.980			20.895		Station D
		0.885		20.990		B.M. 20.99 A.O.D.

2.41   1.950   20.530

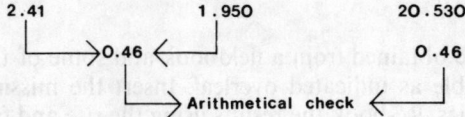

Fig. 8.6 The height of collimation method

Levelling

collimation to be determined. This procedure is outlined below for Station C in Fig. 8.6, where:

1. The height of collimation at instrument position 1
   = reduced level at the bench mark plus staff reading.
   = 20.530 + 1.155
   = <u>21.685 m</u>

2. The reduced level at Station C
   = Height of collimation − foresight at Station C
   = 21.685 − 1.065
   = <u>20.620 m</u>

3. The new height of collimation after moving the instrument can then be determined from the backsight on Station C in the same manner as the first height of collimation was determined, since,

   New height of collimation = reduced level at Station C
                                              + backsight on Station C.
                                              = 20.620 + 1.255
                                              = <u>21.875 m</u>

4. Using the new height of collimation the remaining reduced levels can be determined, remembering that a new height of collimation must be determined for each instrument position.

The arithmetical checks applied to this method of booking exclude the intermediate sights since:

The difference between   The difference between
the last and first   = the sum of the backsights
reduced level   and the sum of the foresights.

While the reduction procedures for this method are often considered to be easier, the system does not check the intermediate sights and a larger number of intermediate sights are often obtained with construction levelling.

A lengthy check which may be applied to this system of booking to include the intermediate sights states:

The sum of each collimation height multiplied by the number of reduced levels obtained from it is equal to the sum of all the intermediate sights, foresights and reduced levels excluding the first reduced level.

Example 8.1

The following entries were obtained from a fieldbook and some of the entries had become illegible as indicated overleaf. Insert the missing entries and check the entries. Re-book the results using the rise and fall method.   (RICS)

Inverted staff readings

Backsight	Intersight	Foresight	Height of instrument	Reduced level	Remarks
?			28.97	27.72	OBM 27.72 m
	2.02			?	Station A
	?			27.94	Station B
1.37		0.98	?	27.99	Station C
	2.16			27.20	Station D
		?		27.89	OBM 27.89 m

The first backsight = height of instrument − Reduced level
$$= 28.97 - 27.72 = \underline{1.25\,\text{m}}$$

Reduced level at Station A = height of instrument − staff reading
$$= 28.97 - 2.02 = 26.95\,\text{m}$$

Height of instrument = reduced level + New backsight
$$= 27.99 + 1.37 = 29.36$$

Last foresight = height of instrument − reduced level
$$= 29.96 - 27.89 = 1.47\,\text{m}$$

Completed fieldbook by rise and fall method:

Backsight	Intersight	Foresight	Rise	Fall	Reduced level	Remarks
1.25					27.72	OBM 27.72 m
	2.02			0.77	26.95	Station A
	1.03		0.99		27.94	Station B
1.37		0.98	0.05		27.99	Station C
	2.16			0.79	27.20	Station D
		1.47	0.69		27.89	OBM 27.89 m
2.62		2.45	1.73	1.56	27.72	
2.45				1.56		
0.17			0.17		0.17	

All the arithmetic checks are correct.

8.5 INVERTED STAFF READINGS

All the levels outlined in this chapter have been observed below the height of collimation, but in certain cases, eg bridge heights, door heads, the transfer of levels over physical obstructions to the line of sight, etc, the vertical distance **above** the line of collimation may be observed. This procedure of observing the distance above the line of collimation is

Levelling

termed an inverted staff reading, and the procedure is illustrated in Fig. 8.7.

$$\text{The height of the staff} = \text{reduced level of Station A}$$
$$+ \text{inverted staff reading}$$
$$= 24.97 - 1.27 - 1.39$$
$$= 27.75 \, \text{m}$$

When booking an inverted staff reading in the fieldbook, a note is made in the remarks column to identify where the reading was located and a **negative** sign is inserted alongside the value. The negative sign will then allow the conventional reduction procedures to be undertaken, since:

$$\text{Reduced level} = \text{height of collimation} - \text{staff reading}$$
$$= 26.24 - (-1.39)$$
$$= 27.75 \, \text{m}$$

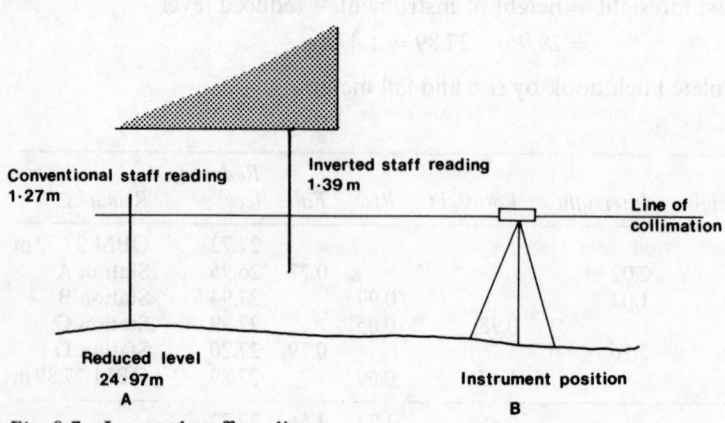

Fig. 8.7 Inverted staff reading

8.6 SOURCES OF ERROR WHEN LEVELLING

Errors will always occur when levelling and it is important that the surveyor;

(a) identifies the permissible error for the survey application;
(b) selects equipment consistent with the required accuracy demanded by the application;
(c) adopts procedures which check each stage of the work and enables corrections to be made when an error is identified.

An error of ± 1 mm in 15 m of distance levelled is usually considered satisfactory for careful work on flat ground or $\pm 20\sqrt{K}$ mm where K is the distance levelled in kilometres over a large distance. The majority of small survey applications do not include such large distances and the error is more practically related to the number of times the instrument is **set up**. Since the majority of errors may occur while undertaking temporary adjustments (section 7.3) an error of $\pm 5\sqrt{N}$ mm where N = number of stations may be considered more satisfactory.

Where a closing error is outside the permissible limits of accuracy then the procedures must be repeated if an inspection of the fieldbook does not identify the source of the error. If the error is within the permissible limits of the survey then the level book should be adjusted so that the error is distributed throughout the **stations** of the survey. Since an error on an intermediate sight will only affect that reading these are not adjusted. The total error will be divided by the number of backsight and foresight readings and the result added or subtracted from each reading, depending on the direction of the error.

For more accurate work the levelling procedures may be repeated and the mean reading (providing no gross error is detected) used for reducing the results.

The basic errors possible when levelling are:

8.6.1 Instrument errors

These may be considered under the following headings:
1. Incorrect setting up may result in
   (a) the bubble off centre when a reading was taken;
   (b) parallax due to incorrect adjustment of the telescope;
   (c) instrument not secure on tripod;
   (d) staff not fully extended.
2. Permanent adjustment of the instrument may be faulty, causing a collimation error. This will be cancelled out if the distance from the instrument to a backsight and foresight are equal. Where it is impossible to maintain equal backsights and foresights **reciprocal levelling** may be undertaken as illustrated in Fig. 8.8.

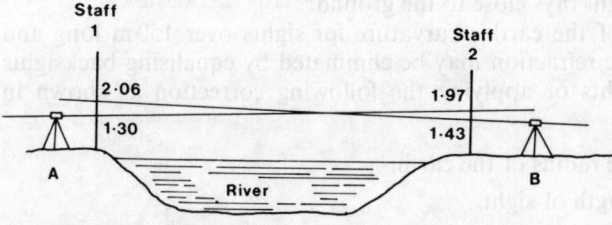

Fig. 8.8 Reciprocal levelling

Levelling

If there is any error in the reading then the mean value will give the true result, since the error from each set of readings should be equal but opposite in sign. This procedure will eliminate **collimation** errors if only one instrument is used. Where two instruments are used, they must be interchanged and the procedure repeated.

Difference in reading at location A = 1.970 − 1.300 = 0.670 m.

Difference in reading at location B = 2.060 − 1.430 = 1.630 m.

True difference in level $= \dfrac{0.630 + 0.670}{2} = 0.650$ m.

The number of readings taken will depend on the accuracy required.

3. Staff errors may occur if the staff is not held vertical (Example 7.1) or the graduations on the staff are not accurate.

8.6.2 Personal errors

These should not occur if correct procedures are followed and the work is checked at each stage of the survey. They include the following

1. The staff is not held vertical. This can be checked with the vertical cross-hair of the instrument and the use of a staff bubble.
2. Movement of the staff occurs during the change over of the level.
3. The bubble is off centre when a reading is taken.
4. The staff is not properly extended.
5. Staff held on soft ground. This effect may be reduced if a change plate is used (Fig. 7.8).
6. Incorrect reading of the staff can be detected by a number of observations at each staff position. This may result from either a misinterpretation of the staff graduations or mistaking a stadia hair for a central hair.
7. Booking errors may occur from a misinterpretation of a reading or a booking in the wrong column. This will be eliminated if all bookings are checked when a second reading is observed.
8. The instrument is not levelled correctly or the tripod is not firm when the reading is taken.

8.6.3 Natural errors

1. The effect of high wind or heat shimmer can be reduced by shading or avoiding sight rays close to the ground.
2. The effect of the earth's curvature for sights over 150 m long and atmospheric refraction may be eliminated by equalising backsights and foresights or applying the following correction as shown in Fig. 8.9.

If R = the radius of the earth.

D = length of sight.

S = error at distance D km.

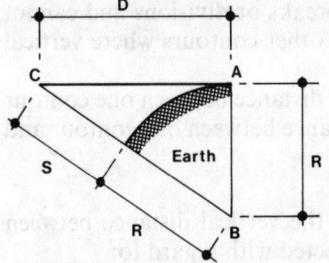

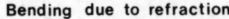

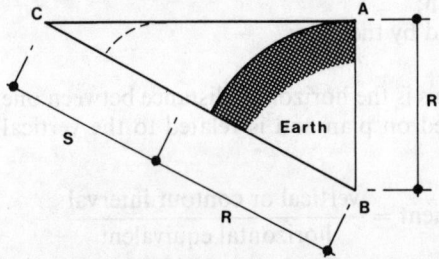

Fig. 8.9 Curvature and refraction

in the right-angled triangle ABC

$$(R + S)^2 = R^2 + D^2$$
$$R^2 + 2RS + S^2 = R^2 + D^2$$

Since S is negligible in relation to R and D

$$S = D^2/2R$$
$$= -0.0785 D^2 \text{ m} \quad \text{(if R = 6370 km and D is in kilometres)}$$

Atmospheric refraction causes the line of sight to bend downwards towards the horizon and a correction of 1/7, the curvature correction, may be assumed but in the opposite direction. The total correction for curvature and refraction will be:

$0.0675 d^2$ m where d is in kilometres

This value will be subtracted from the staff reading.
3. Where bad light is experienced the length of sights should be reduced.
4. Raindrops on the object lens can be reduced by use of the shade.

8.7 APPLICATIONS OF LEVELLING

8.7.1 Contouring

A contour is a line on a map which connects points of equal altitude and is shown in a distinct colour to enable the topography to be assessed.

Levelling

These lines will be continuous without breaks or divisions and cannot cross each other although they may join other contours where vertical cliffs are indicated.

Where the ground slope is steepest the distance between one contour and the next will be a minimum. The distance between one contour and the next may be expressed as:

The vertical or contour interval. This is the vertical distance between one contour and the next and will be selected with regard to:

(a) the application of the survey;
(b) the scale of the map or plan;
(c) the degree of slope enclosed by the map;

The horizontal equivalent. This is the horizontal distance between one contour and the next measured on plan and is related to the vertical interval, since:

$$\text{Ground slope or gradient} = \frac{\text{vertical or contour interval}}{\text{horizontal equivalent}}$$

The following methods of contouring are commonly used:

Direct contouring. The level is first set up at some convenient point and the height of collimation determined from a backsight on to a datum of known level. The surveyor will then direct the staff man about the field to identify the positions where the staff readings are the same. Assuming the height of collimation is 101.9 m; then all the locations where a staff reading of 1.9 m are identified will locate the 100 m contour and this is repeated for other selected contours.

Each contour may be located by a different coloured peg which is then located by conventional chain surveying procedures for plotting on the finished plan.

This procedure has advantages in hilly terrain since it is the most accurate procedure available and is not dependent on associated calculations when plotting. The procedure is limited however, since it is slow and requires extra chaining to locate the pegs, which make it costly.

The grid method. This is an ideal procedure for relatively small flat sites where a square grid may be set out using a chain or a theodolite. The grid interval will be determined by the nature of the ground and the level of detail required but an interval of 10 m to 20 m is usually selected. Levels are then recorded at each intersection of the grid with a note in the remarks column of the fieldbook to indicate the grid reference. Only two sets of adjacent sides of the grid need setting out in the field since the remaining intersections can be located with ranging rods or one set where two theodolites are available. The position of selected contours

Applications of levelling

can then be determined by an interpolation graph, illustrated in Fig. 8.10 which is drawn on tracing paper and placed over the plotted grid, or by calculation.

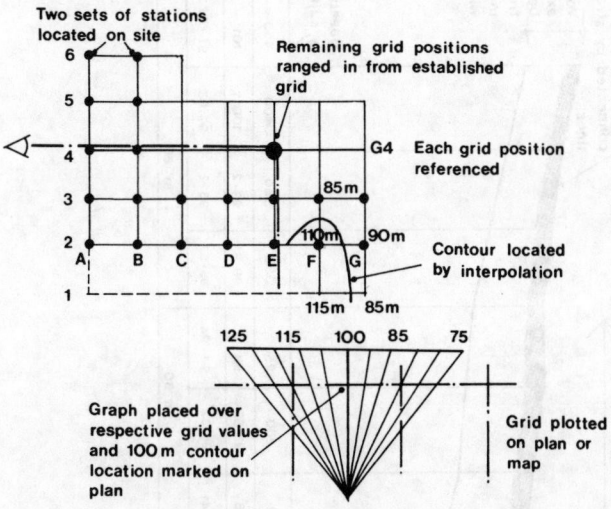

Fig. 8.10 The grid method of contouring

When calculation procedures are used the following calculation sheet may be used to formalise the work.

Stations	Fall or rise	Contour fall or rise	Distance	From–to
F1–G1	115 − 85 = 30 m	115 − 100 = 15	$\frac{15}{30} \times 20 = 10$	F1–G1
F2–G2	110 − 90 = 20 m	110 − 100 = 10	$\frac{10}{20} \times 20 = 10$	F2–G2

Note that each calculation is relative to the first station on the line identified and in the example shown all calculations are related to a 20 m grid and the 100 m contour.

The procedure assumes that the ground slope is uniform, which reduces the accuracy, although it is faster on site and allows the average level of the grid area to be determined since:

Average level of grid
$$= \frac{\text{Sum of reduced levels at each grid position} \times \text{number of squares meeting at that position}}{4 \times \text{the number of squares contained by the grid}}$$

Levelling

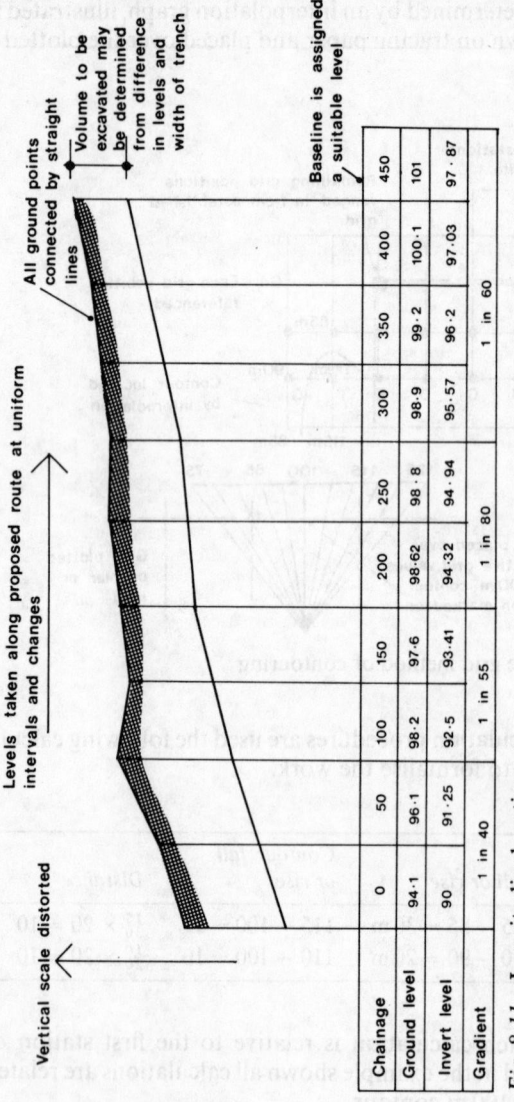

Fig. 8.11 Longitudinal sections

Contours for calculating volumes. The volume of earth enclosed by a particular area can be determined by tracing the path of each contour with the tracer of a planimeter. This will then give the area enclosed by a particular contour which may be multiplied by the contour interval to identify the volume (section 11.10).

Longitudinal sections. These are often required for construction work when road works demand that the excavated soil should balance the fill or where drains have to be laid to a given fall. The sections are plotted to a distorted scale in order to emphasise the profile and a scale of 1/500 may be used horizontally while 1/100 would be satisfactory for the vertical scale.

The volume of excavation required for a pipeline, etc can be calculated from the longitudinal section when levels have been taken at uniform intervals and these procedures are described in (section 11.8).

8.8 EXERCISES

1. The following readings were taken as part of a levelling traverse:

Backsight	0.871 on to a bench mark of reduced level of 19.74 m AOD.
Foresight	1.652
Backsight	2.327
Intermediate sight	1.635, 1.104, 2.431, 0.169 on to pegs numbered 820, 821, 822 and 823 respectively.
Foresight	0.786
Backsight	2.243
Foresight	1.215 on to a temporary bench mark.

   Book the readings evaluated all reduced levels and carrying out all standard checks on arithmetic by

   (a) rise and fall and
   (b) collimation methods of booking. (RICS)

2. State fully what is meant by:

   (a) bench mark;
   (b) line of collimation;
   (c) backsight;
   (d) parallax.

   Illustrate your answers with sketches. (RICS)

3. The readings taken on a levelling traverse are as follows:

   Backsight on to a bench mark of reduced level 101.630 m is 3.421

Levelling

Foresight 2.427
Backsight 3.210
Intermediate sight on to a manhole cover 2.193
Intermediate sight on to a gateway 1.074
Foresight 0.231
Backsight 2.116
Intermediate sights 0.321 proposed sight of manhole G
                 0.864 at 10 m from G on line of sewer
                 1.369 at 20 m
Foresight 2.160 at 30 m
Backsight 1.403
Intermediate sights 1.673 at 40 m
                 2.229 at 50 m
                 3.406 at 60 m
Foresight 3.503 at 65 m sight of manhole F.
Book these readings (a) by rise and fall
                      (b) by collimation

9 TACHEOMETRY

9.1 ADVANTAGES AND LIMITATIONS

This is a method of measuring both horizontal and vertical distances between known points by means of a staff using the optical properties of a telescope. Since conventional tapes, etc are not required for measuring linear distances the procedure has the following advantages:

(a) observations are not affected by the nature of the terrain;
(b) the procedures are simple to undertake;
(c) access may not be required on to property;
(d) it is usually less time consuming;
(e) gross errors are readily identified;
(f) Calculations are absent in many cases;
(g) obstructions at ground level do not hinder the work.

However, the following limitations must be considered:

1. The accuracy can vary between 1:500 up to 1:10 000 and should only be used at the lower accuracy for filling in the detail of a survey not for establishing a control network.
2. The equipment is more expensive in many cases, especially for the higher accuracy work.

All the systems used are derived from the properties of an isosceles or right-angled triangle where the apex angle referred to as the 'parallactic' angle is measured, along with the length of one side of the triangle (see Fig. 9.1). The method of forming this triangle will determine the procedure being adopted, and two basic systems are available:

1. Vertical staff tacheometry.
2. Horizontal staff tacheometry.

Tacheometry

9.2 VERTICAL STAFF TACHEOMETRY

Two methods of observation may be undertaken, depending on whether the parallactic angle is fixed or variable when the reading is observed from a staff held **vertical** over a station.

9.2.1 Tangential tacheometry

Horizontal sights. This is the simplest form of tacheometry and possibly the least accurate. The telescope is only required to have a horizontal cross-hair and a standard staff can be used in the following manner:

(a) set up the telescope at Station A and note the height of the line of collimation above ground level;
(b) set the telescope horizontal and sight on to the staff at Station B;
(c) incline the telescope, noting the angle of elevation and take a further reading on to the staff;
(d) note the angle of elevation and the difference between the staff readings.

Example 9.1

A telescope is located 1.2 m above Station A and a reading of 2.622 m is noted on a staff at Station B when the telescope is horizontal. The telescope is elevated through an angle of 9° 20′ and a further reading of 3.762 m is noted (Fig. 9.1). Calculate the distance between Stations A and B.

Since the triangle EFG is a right-angled triangle, with GF being the horizontal distance between A and B:

$$\frac{EF}{GF} = \tan 9° 20'$$

$$GF = EF/\tan 9° 20' \text{ m}$$

$$GF = 1.14/\tan 9° 20' \text{ m}$$

$$GF = \underline{6.936 \text{ m}}$$

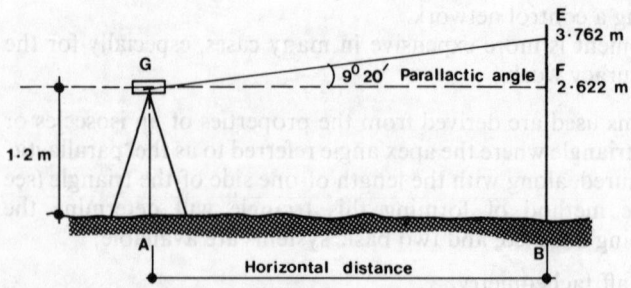

Fig. 9.1 Tangential tacheometry

Vertical staff tacheometry

The horizontal distance between AB is 6.94 m.

This procedure can be speedily undertaken on a modern calculator as the observations are made.

If the reduced level of Station A was known to be 98.425 m (AOD) then the reduced level of Station B is:

$$98.425 \text{ m} + (1.2 \text{ m} - 2.622 \text{ m})$$
$$98.425 \text{ m} - 1.422 \text{ m}$$
$$\underline{97.003 \text{ m AOD}}$$

Inclined sights. In a situation where B is above the line of collimation of the instrument (Fig. 9.2), the following procedure may be adopted:

$$\frac{df}{Gd} = \tan \theta°$$

$$df = gd \cdot \tan \theta°$$

$$de = Gd \tan \alpha°$$

$$S = Gd(\tan \alpha° - \tan \theta°)$$

$$Gd = S/\tan \alpha° - \tan \theta°$$

To obtain the differences in height between the stations.

$$fd/Gd = \tan \theta$$
$$Bd = df - fB$$
$$= Gd \tan \theta° - Bf$$
Difference in level = GA + Bd = V
$$V = GA + (Gd \tan \theta - Bf) \text{ m}$$

Since $Gd \tan \theta°$ has already been obtained and Bf is the staff intercept reading, the difference in level can be determined.

Example 9.2

In Fig. 9.2 assume:

Angle DGF = 6° 30' 20" staff reading f = 1.675 m
Angle DGE = 6° 58' 40" staff reading h = 3.980 m

The instrument height at Station A is 1.290 m, which is 52.360 m (AOD).

Calculate the plan length between Stations A and B.

$$S = 3.980 - 3.675 \text{ m}$$
$$= 0.305 \text{ m}$$
$$H = S/\tan \alpha° - \tan \theta°$$
$$H = 0.305/0.008 \text{ m}$$
$$\underline{\text{Horizontal distance } 36.496 \text{ m}}$$

Tacheometry

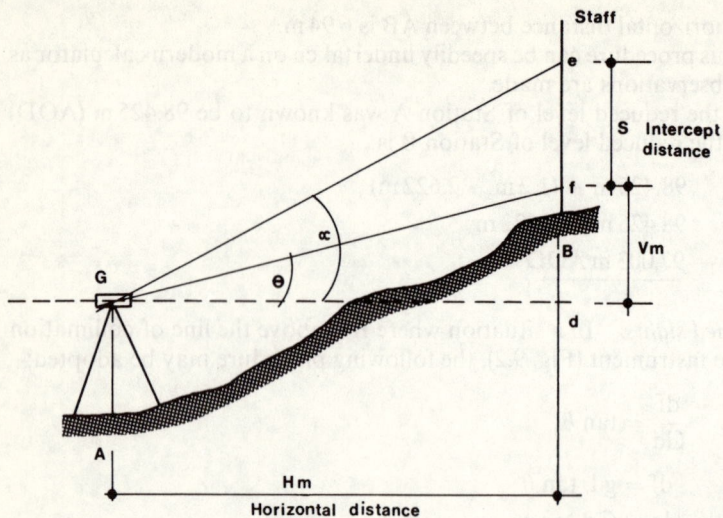

Fig. 9.2 Inclined staff readings

Note that the staff intercept distance for this procedure is normally small, which reduces the accuracy of the solution and limits its application.

Methods of reducing calculations. The above calculations may be simplified by the following procedures, although they may increase the survey period in the field due to the sighting intercepts being fixed:

(i) fix the staff intercept distance at 2 m and fit targets on the staff at this distance to make sighting easier;
(ii) fix the value of $\tan \alpha° - \tan \theta°$ at 0.01. This will require $\tan \theta$ to be calculated, once angle α has been observed in the field.

If is 10° 20′ then $\tan \theta = 0.1823$

$$\tan = 0.1723 = (0.1823 - 0.01)$$
$$= 9°46′34″$$

S may then be determined

9.2.2 Stadia tacheometry

Principles. This is the commonest form of vertical staff tacheometry today since the theodolite incorporates three stadia hairs on the graticule as indicated in Fig. 9.3. The triangle defined by sighting along each stadia hair will allow the plan distance and vertical height to be

Vertical staff tacheometry

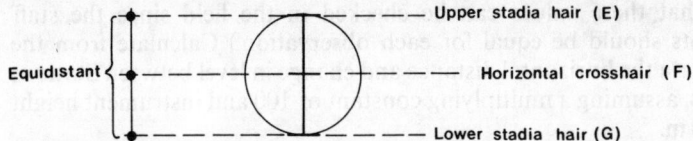

Fig. 9.3 View through the eyepiece of a telescope

determined as shown in Fig. 9.4. From similar triangles:

$$\frac{d}{f} = \frac{s}{i}$$

$$d = s.f/i$$

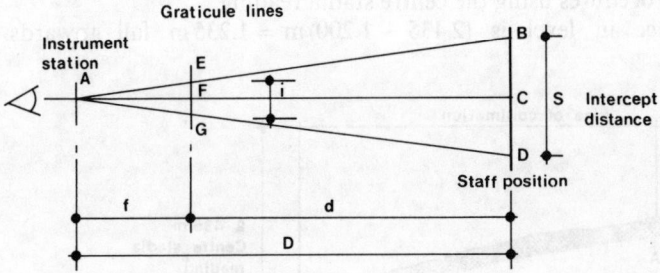

Fig. 9.4 Stadia arrangement

Since f = the distance of the graticule lines from the eyepiece
i = the distance between upper and lower stadia hair.

Both of these dimensions are a property of the telescope and may be fixed by the manufacturer so that:

f/i, which is called the **multiplying constant**, can be given a value of **100** in modern instruments.

By fixing the multiplying constant at 100, the surveyor has only to **multiply** the staff intercept distance by 100 to obtain the plan distance under consideration, eliminating the need for trigonometrical tables as required for tangential tacheometry.

Example 9.3

A theodolite is set up over Station A and sighted on to a staff at Station B to give the following observations:

Upper stadia reading = 2.735 m.
Centre stadia reading = 2.435 m.
Lower stadia reading = 2.135 m.

Tacheometry

(Note that these values can be checked in the field since the staff intercepts should be equal for each observation.) Calculate from the above data the horizontal distance and change in level between Stations A and B, assuming a multiplying constant of 100 and instrument height of 1.200 m.

First check the data

$$(2.735 - 2.435) = (2.435 - 2.135)$$
$$\underline{0.300 = 0.300} \quad \text{(data is correct)}.$$

Difference between upper and lower stadia readings, 0.600 m.

Horizontal distance = $0.600 \times 100 = \underline{60.00 \text{ m}}$

The difference in level (Fig. 9.5) may be determined by ordinary levelling procedures using the centre stadia reading.

Difference in level is $(2.435 - 1.200) \text{ m} = 1.235 \text{ m}$ fall towards Station B.

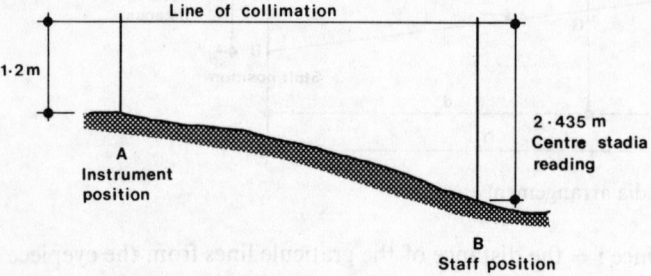

Fig. 9.5 Difference in level

Adjustments for internal focusing telescopes. The basic formulae D = MS is only true for an external focusing telescope where the focal length is constant. Modern telescopes adopt an internal focusing mechanism, which requires the following correction to be made, since the focal point of the lens system and the vertical axis of the instrument may not coincide, as indicated in Fig. 9.6.

The simple optical arrangement in Fig. 9.6 changes the basic stadia geometry, since:

$$d/f = s/i$$
$$d = s \times f/i$$

But the horizontal distance D is required, therefore:

$$D = d + c$$
$$D = (s \times f/i) + c$$

c being the **additive** constant.

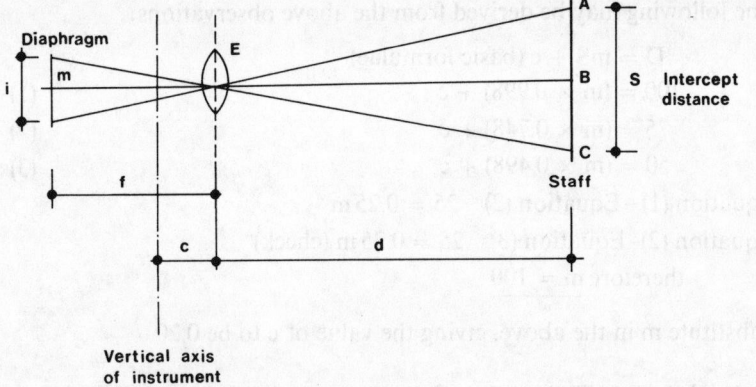

Fig. 9.6 The optical arrangement of the internal focusing telescope

This gives a basic formulae for the distance of

$$D = s \times f/i + c$$

The additive constant (c) in modern telescopes is usually very small and is often given a value of zero. Alternatively, a convex lens could be fitted within the telescope relative to the object lens, so as to form the image at the correct position. This lens is termed an anallactic lens but is seldom found in modern instruments due to the small size of the error that is likely when the additive constant is ignored.

Determination of constants by field observation. The multiplying constant (m) and the additive constant (c) may be determined in the field by levelling the instrument in the field and sighting on to three stations 25 m, 50 m and 75 m distant to obtain the following results:

Distance	Stadia reading			Staff Intercept distance
	Top	Middle	Bot	
100	1.952	1.4533	0.954	0.998
75	1.670	1.296	0.922	0.748
50	1.342	1.093	0.844	0.498

The general formulae for the horizontal distance is given by:

$$D = mS + c$$

where:

D = the horizontal distance
S = staff intercept distance
m = multiplying constant
c = additive constant

Tacheometry

The following may be derived from the above observations:

$$D = mS + c \text{ (basic formulae)}$$

$$100 = (m \times 0.998) + c \quad (1)$$
$$75 = (m \times 0.748) + c \quad (2)$$
$$50 = (m \times 0.498) + c \quad (3)$$

Equation (1) – Equation (2) $25 = 0.25\,m$
Equation (2) – Equation (3) $25 = 0.25\,m$ (check)
therefore $m = \underline{100}$

Substitute m in the above, giving the value of c to be 0.20.

Inclined sights. The previous formulae were derived for an instrument adjusted to the horizontal plane but on steep ground, as indicated in Fig. 9.7, this may not be possible. Since the line of sight is reduced due to the ground contours when the instrument is horizontal the telescope has been elevated by $\theta°$ to sight on to Station B. This will result in OC being the distance calculated instead of the desired horizontal distance, although this will contain an error, since $D_1 E_1$ is the intercept distance required and not DE, which is greater.

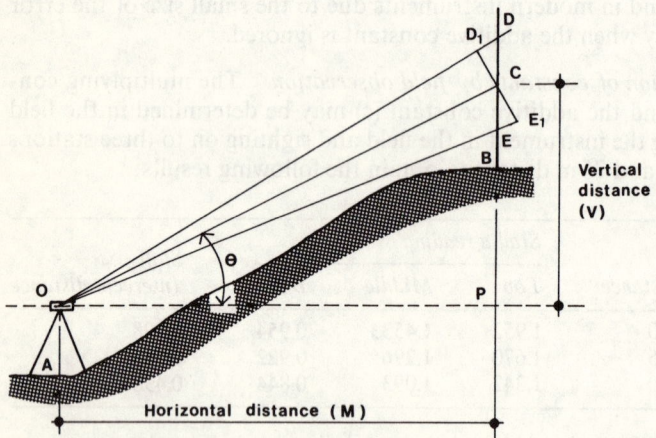

Fig. 9.7 Arrangement for inclined sights

Staff inclined. DE may be determined if the staff is inclined at right angles to the line of sight for each reading and the following data obtained:

Since: $D = mS + c$
         $OC = mS + c$

Vertical staff tacheometry

In triangle OPC:

Since: $OP/OC = \cos\theta°$
$$OP = OC\cos\theta° = (mS + k)\cos\theta°$$
$$= \underline{\text{Horizontal distance}}$$

Vertical distance $= CP = OC \sin\theta°$
$$CP = (mS + k)\sin\theta°$$
$$= \underline{OP \tan\theta°}$$

Whilst care must be taken to incline the staff for each reading, the procedure may be used for the occasional steep sight in either elevation or depression and a device is available for attaching to the staff to note the actual inclination of the staff which must be set to the **vertical** circle reading of the theodolite.

Staff vertical. For angles of elevation or depression up to 25° the staff is normally held vertical and the following adjustments made to the reading (Fig. 9.7).

Whilst D_1E_1 is the required staff intercept distance, the reading observed will be DE, which is greater due to the staff being vertical, and the reading must be corrected as follows.

In triangle DD_1C, $\angle D_1\hat{C}D = \theta°$
$\angle D\hat{D}_1C$ may be considered a right angle.

then

$D_1C/DC = \cos\theta°$
$\therefore\quad D_1C = DC \cos\theta°$
$\therefore\quad D_1E_1 = DE \cos\theta°$

From the basic formulae:

$$D = (s \times f/i) + C \text{ when } f/i = m$$
$$OC = (M \times D_1E_1) + C$$
$$OC = (M \times DE\cos\theta°) + C \quad (\text{since } D_1E_1 = DE\cos\theta°)$$

In triangle OCP:

Horizontal distance $\quad OP = OC \cos\theta°$
Vertical distance $\quad PC = OC \sin\theta°$
$\therefore\quad OP = ((M \times DE\cos\theta) + C) \cos\theta°$
$\qquad = \underline{M.DE \cos^2\theta° + C\cos\theta°}$

$\therefore\quad CP = [(M \times DE\cos\theta) + C]\sin\theta°$
$\qquad = M.DE \cos\theta \sin\theta \sin\theta + C\sin\theta°$

Tacheometry

Since θ will be a small angle:

$\cos \theta° \simeq 1$
$\sin \theta° \simeq 0$

Therefore the above formulae for vertical staff readings are reduced to:

Horizontal distance = $MS \cos^2 \theta° + C$
Vertical distance = $MS \cos \theta° \sin \theta°$

where

M = the multiplying constant
S = staff intercept distance
C = the additive constant

If we reconsider triangle OPC

$\tan \theta° = CP/OP$
$OP \tan \theta° = CP$

Vertical distance = Horizontal distance × $\tan \theta°$

Example 9.4

A theodolite sights on to a BM 25 m AOD with an angle of depression of 10°. Determine the height of the instrument if stadia readings are 2.7 m, 1.6 m, 0.5 m.

If the same instrument now sights on to a station with an angle of depression of 15° and the stadia readings are 2.5 m, 1.6 m, 0.7 m, find the distance of the staff station and its reduced level. (RICS)

$V = f/i \times S \cos \theta . \sin \theta$ (basic formula)
$V = 100(2.7 - 0.5) \sin 10° . \cos 10°$
$ = 100 \times 2.2 \times 0.1710$
$ = \underline{37.62 \, m}$

Height of instrument = BM + V + h
$ = 25 + 37.61 + 1.6$
$ = \underline{64.21 \, m}$

$H = (f/i \times s \times \cos^2 15°) + C$
$ = (100 \times 1.8 \times 0.9659^2) + 1.0$
$ = \underline{168.93 \, m}$

$V = f/i . s . \cos \theta° \times \sin \theta°$
$ = 100 \times 1.8 \times 0.9659 \times 0.2588$
$ = \underline{45 \, m}$

Reduced level = height of instrument − V − h
= 64.21 − 45 − 1.60
= 17.61 m

Example 9.5

The following data was taken between two stations with a tacheometer with the staff held vertical, the multiplying constant being 100 and the additive constant 0.

Staff position	Horizontal circle	Vertical circle	Staff readings		
			Lower	Middle	Upper
A	10° 40′	3° 24′	1.630	1.935	2.240
B	92° 46′	−6° 14′	2.040	2.241	2.442

Determine the gradient between Stations A and B.   (RICS)

At Station A

$H = (f/i) \times s \cdot \cos^2 \theta° + C$
$= 100 \times 0.61 \times \cos^2 3° 24'$
$= \underline{60.78 \text{ m}}$

$V = + (f/i) \cdot s \cdot \cos \theta \sin \theta$
$= 100 \times 0.61 \times 0.9999 \times 0.0593$
$= \underline{+3.617 \text{ m}}$

Note that the positive sign for V indicates an observation above horizontal.

At Station B

$H = (f/i) \times s \cdot \cos^2 \theta + C$
$= 100 \times 0.402 \times \cos 6° 14'$
$= \underline{39.726 \text{ m}}$

$V = (f/i) \times s \cdot \cos \theta \sin \theta$
$= 100 \times 0.402 \times \cos 6° 14' \times \sin 6° 14'$
$= \underline{-4.338 \text{ m}}$

The difference in level between Stations A and B

$= (3.617 − 1.935) + (4.338 − 2.2241)$
$= \underline{8.261 \text{ m}}$

Tacheometry

The distance between A and B may be determined from the horizontal angle observations using the cosine rule:

$$AB^2 = 60.78^2 + 39.726^2 - 2(60.78 \times 39.726) \times \cos 82° 06'$$
$$AB = \sqrt{5244.735}$$
$$= \underline{72.42\,m}$$

The gradient between AB = 8.261/72.42
$$= \underline{1/8.766}$$

9.2.3 Errors in vertical staff tacheometry.

Gross errors. These may result from:

(a) wrong reading of the staff;
(b) wrong reading of the vertical angle;
(c) wrong booking;
(d) error in calculation.

Since there are many observations to book with stadia tacheometry it is important that the observer does not also act as booker, in order to reduce the possibility of gross errors. The booker should be able to reduce the gross errors by checking that the difference between the lower and centre reading of the staff equals the difference between the upper and centre staff reading. The correct procedures must be followed and each stage checked to reduce the above errors.

Systematic errors. These may result from:

(a) non-perpendicularity of the staff providing the wrong staff readings, ie $A_1B_1 = AB(1 + \cos \alpha°)$ in Fig. 9.8;
(b) non-perpendicularity of the staff providing the wrong horizontal distance

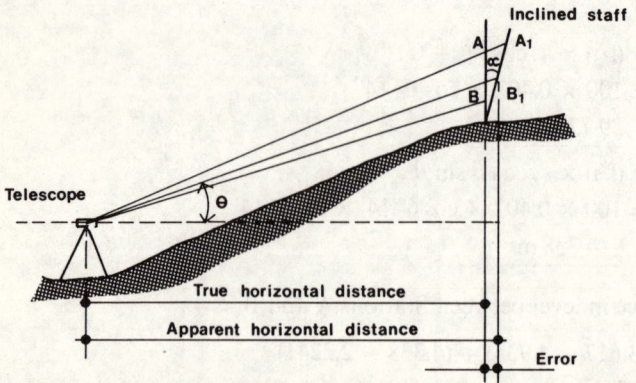

Fig. 9.8 Inclined staff error

Vertical staff tacheometry

Table 9.1 Example of fieldbook entry

Title: Date:										Instrument no. Mult. constant Add. constant	
Staff stn.	Height of inst.	Horz. angle	Vert. angle	Stadia reading	Inter dist.	Horz. dist.	Vert. dist.	Rise + fall −	RL Inst. axis.	RL Staff stn.
Instrument station A	1.75	0	3° 18′	2.125 1.750 1.375	0.75	74.75	4.31	+2.56	12.44	15 m TBM
X		31° 42′	0°	3.060 2.850 2.640	0.42	42.00	0	−2.850		9.59
B		38° 20′	−0° 36′	3.210 2.840 2.470	0.74	73.99	−0.77	−3.61		8.83
Instrument station B										
A		200° 15′	0° 54′	0.970 0.580 0.090	0.88	87.97	1.38	+0.85	5.63	

Completed in the field ⟵⟶ Computations carried out later

Tacheometry

If the telescope in Fig. 9.8 has no additive constant:

Apparent distance = $M \times A_1 B_1 \cos^2 \theta$
Actual distance $= M \times A \times B \cos^2 \theta$

Error = Apparent distance − Actual distance

$$= 1 - \frac{\cos(\theta + \alpha)}{\cos \theta}$$

A level should be attached to the staff in order to reduce both of the above errors.

Differential refraction. This happens when working being close to the ground, and can be reduced on hot days by working at night, with illuminated targets when a high order of accuracy is demanded.

9.2.4 Booking procedures for vertical staff tacheometry

Table 9.1 illustrates a booking format which enables the field entries to be input and validated while the reduction calculations can be undertaken later. Before the field observations are reduced the following data checks **must** be undertaken and any errors corrected:

(a) all the data has been completed for each reading;
(b) the survey has been referenced to a point of known height and orientation (these may be separate stations);
(c) the difference between the upper and centre stadia reading is equal to the difference between the centre and lower stadia reading for each observation;
(d) at each change of instrument position the stadia readings and bearings have been noted and referenced to **both** stations, eg observations from Station A to Station B are made before moving the instrument to Station B and after occupying Station B the first reading will be **back** to Station A.

If the above checks on the data are satisfactory the following calculations may be undertaken:

1. Calculate the staff intercept distance of each observation.
2. Calculate the **horizontal** and **vertical** distances from the instrument to the staff position using the respective formulae (page 143).
3. Determine the **rise** or **fall** between the centre stadia reading and the vertical distance V.
4. Calculate the reduced levels of each station.

This procedure is outlined below (Fig. 9.9) using the fieldbook entries illustrated in Table 9.1.

Reduced level of instrument axis = $15\,\text{m} - 2.56$
$= \underline{12.44\,\text{m}}$

Vertical staff tacheometry: tangential tacheometry

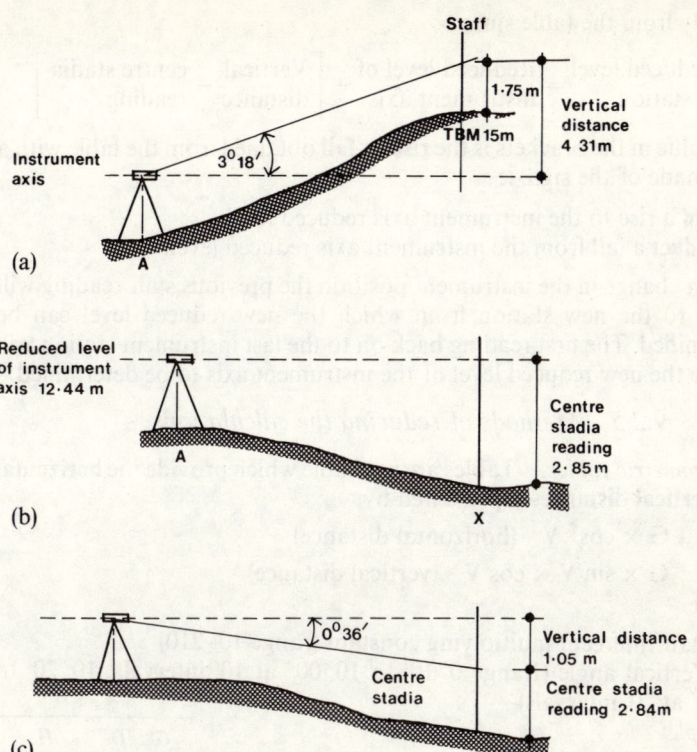

Fig. 9.9 Booking procedure (a) sight on to A (Bench Mark) (b) sight on to station X (c) sight on to station B

The reduced level of the instrument axis is then inserted in the respective column of the booking sheet opposite the instrument station and each observation from that instrument station is reduced using conventional levelling procedures.

The reduced level of Station X = 12.44 − 2.850
$$= \underline{9.59 \text{ m}}$$

(*Note* that the negative sign indicates the reduced level is **below** the instrument axis as illustrated in Fig. 9.9(c)).

The reduced level at Station B = 12.44 − (0.77 − 2.84)
$$= 12.44 - 3.61$$
$$= \underline{8.83 \text{ m}}$$

The illustrations and basic steps outlined above demonstrate the procedures to be used although each reduced level can be obtained

Tacheometry

directly from the table since:

$$\text{The reduced level at the station} = \text{Reduced level of instrument axis} + \left[\text{Vertical distance} - \text{centre stadia reading}\right]$$

The value in the brackets is the **rise** or **fall** obtained from the table with a note made of the sign, ie

(a) **add** a rise to the instrument axis reduced level;
(b) **deduct** a fall from the instrument axis reduced level.

At a change in the instrument position the previous staff reading will be on to the new station from which the new reduced level can be determined. The first reading back on to the last instrument station will enable the new reduced level of the instrument axis to be determined.

9.2.5 Methods of reducing the calculations

Tacheometric tables. Tables are available which provide the **horizontal** and **vertical** distances represented by:

$G \times \cos^2 V$ (horizontal distance)

$G \times \sin V \times \cos V$ (vertical distance)

Where

G – Staff intercept multiplying constant (range 10–210)
V – Vertical angle (Range 0° 10′ to 10° 00″ at 10′ intervals, 10° 20′ to 20° 00′ at 20′ intervals)

Example:

Vertical angle = 4° 50′
Staff intercept = 0.45 m
= 45

By inspection of the tacheometric table for 4° 20′;

Horizontal distance = 44.68 m
Vertical distance = 3.78 m

G	D	H
43	42.69	3.61
44	43.68	3.78
45	44.68	3.78
46	45.67	3.86
47	46.67	3.95

(From D. T. F. Munsey, *Tacheometric Tables*, Technical Press, 1971)

Electronic calculators. The application of programmable calculators to the reduction of the results can be used since they are capable of undertaking all the necessary calculations for reducing $\cos^2 0$ and $\cos 0 \cdot \text{sine}$ for each observation. The procedure followed will depend on the following:

(a) language used for programming;
(b) function keys available;
(c) levels of sub-routine available;
(d) number of program steps available;
(e) number of memories available.

Vertical staff tacheometry: tangential tacheometry

The outline flowchart in Table 9.2 illustrates the steps required to display the **horizontal** and **vertical** distances from input of:

1. Upper stadia reading.
2. Lower stadia reading.
3. Vertical angle.

The procedures will take approximately seventy steps, depending on the functions available on the calculators and will require ten memory allocations. These basic procedures can be extended to include the reduction of levels and the numeric title of each station when a printer is used for the data output.

Table 9.2   Electronic calculator routine

Input data	Memory
Upper stadia reading	M1
Lower stadia reading	M2
Vertical circle reading	M3
Theodolite constant	M4

Calculations	Memory
M1 − M2	M5
Cos (M3)	M6
M6 × M6 × M5	M7
(Sin (M3) × M6) × M5	M8
M8 × 100	M9
(M7 × 100) + M4	M10

Go to 00	
Return for next data input	

Display	Memory
Vertical distance	M9
Horizontal distance	M10

Micro-computers. Modern micro-computers have ample memory capacity to reduce the stadia readings and vertical angles, providing a tabulated list of:

(a) horizontal distance to each staff station;
(b) vertical distance of each observation;
(c) reduced level of each station referenced to a given datum;
(d) gradient between each station.

With the level of graphics provided by the visual display unit, it is possible to include graphical detail, including a diagram of the

Tacheometry

theodolite and staff, to simplify the input procedures for the user. This level of input will indicate the upper, middle and lower stadia positions when input is required along with the vertical and horizontal scales of the theodolite. Printers are capable of producing hard copy of the results, although the conventional printer is not of sufficient accuracy to plot the stations directly. With the larger memory capacity of this equipment when related to the calculator, the input data can be checked and verified by the program before any calculations are undertaken.

Computer services. Bureaux facilities are available with packages to both compute (Fig. 9.10) and plot data obtained from tacheometric surveys. These systems are flexible and provide the following:

(a) a wide range of theodolite systems may be used;
(b) the direct input of field notes;
(c) provision of accurate contour maps;
(d) isometric projections of the survey from a number of angles;
(e) cross-sections;
(f) provision of material volume data.

The above output data will be provided after input of the following field information:

1. Description of traverse.
2. Instrument multiplying constant.
3. Co-ordinates of the reference station.
4. Levels of the staff/instrument at one location on the survey.
5. Stadia readings.
6. Height of theodolite above ground level.
7. With the self-reducing tacheometer direct height and distance values may be input.

The tacheometric staff. Whilst any well graduated levelling staff may be used for tacheometry, special staffs are available which contain a wedge-shaped mark above the base for accurate centring. The numbering on the staff commences above the target, eliminating any need for the subtraction of stadia readings and thus providing a direct reading of the stadia interval. The makers of tacheometric staffs claim they will improve the accuracy of the results by 20 mm to 50 mm over a distance of 70 m. Above this distance refraction problems close to the ground will reduce the accuracy of the procedure (Fig. 9.11).

The self-reducing tacheometer. The self-reducing tacheometer is constructed similarly to the theodolite with vertical and horizontal circles, telescope and tribrach fitting with optical plummet to locate the instrument over the station. The telescope contains a prism which first projects the image on to the diaphragm plate that contains a series of

Back station	Fore station	Horizontal angle			Vertical angle			Height factor	Staff reading			Inst. height	Seq.	Card type	Remarks
		o	/	//	o	/	//		Top	Centre	Bottom				

(a) Computer Input Sheet

Horizontal			*In*		*Partial coordinates*		*Total coordinates*		*Total*	
Line	*Angle*	*Azimuth*	*Distance ft/m*	*Level ft/m*	*Departure ft/m*	*Latitude ft/m*	*Departure ft/m*	*Latitude ft/,*	*Stn*	*Level ft/m*
8–040	196 27 0	109 43 0	335.000	55.04	96.123	−34.448	200226.823	200248.952	040	221.07
8–041	193 26 0	106 42 0	358.000	45.90	104.517	−31.357	200235.217	200252.043	041	211.93
8–042	198 30 0	111 46 0	342.000	53.48	96.811	−38.656	200227.511	200244.744	042	219.51
8–043	192 33 0	105 49 0	366.000	45.09	107.334	−30.406	200238.034	200252.994	043	211.12
8–044	194 58 0	108 14 0	357.000	54.69	103.351	−34.047	200234.051	200249.353	044	220.72
8–045	190 13 0	103 29 0	363.000	33.11	107.594	−25.798	200238.294	200257.602	045	199.14
8–046	189 4 0	102 20 0	353.000	13.38	105.113	−22.982	200235.813	200260.418	046	179.41
8–047	189 2 0	102 18 0	350.000	21.57	104.233	−22.726	200234.933	200260.674	047	187.60

(b) Sample Data Output

Fig. 9.10 Computer data formats (Courtesy Compower, the NCB computer bureau service)

Tacheometry

reduction curves to give the image illustrated in Fig. 9.12 when viewed through the telescope. As the telescope is tilted the diaphragm plate revolves, allowing the **horizontal** and **vertical** distances to be obtained from the observed staff intercept values. The difference between the **zero curve** (the lowest curve) and the **distance curve** (the upper intercept) is multiplied by a factor of 100 to give the **horizontal distance**.

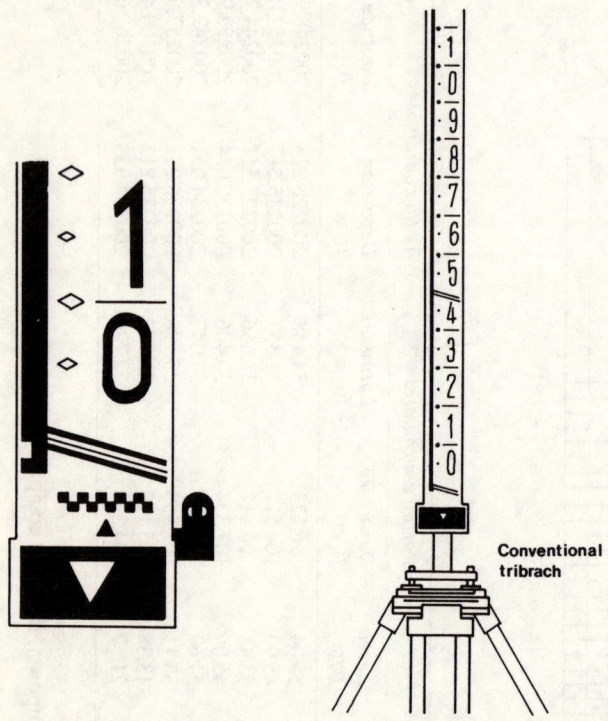

Fig. 9.11 The tacheometric staff (courtesy Wild Heerbrugh)

Example 9.6
In Fig. 9.12, $100(1.17 - 1.00) = 17$
   Horizontal distance = 17 m

The **vertical distance** is obtained by multiplying the difference between the **zero curve** and the **height reading curve** (the curve below the distance curve) by the multiplying factor displayed.

Example 9.7
In Fig. 9.12, $0.1 \times 0.8 = 0.08$ m
   Vertical distance = 0.08 m

Horizontal staff tacheometry

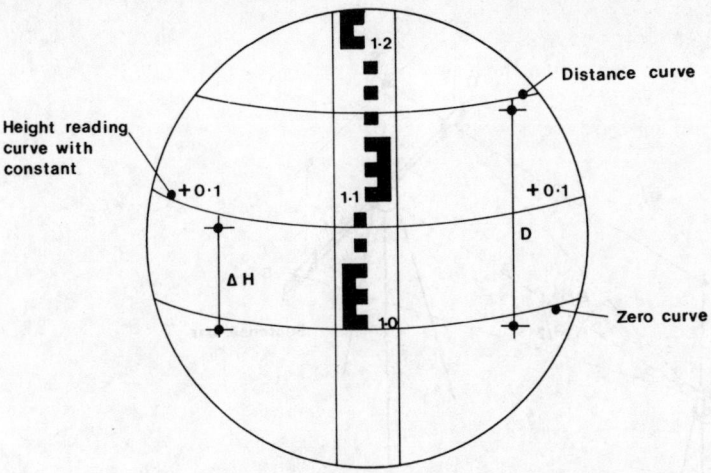

Fig. 9.12 The self-reducing tacheometer (courtesy Wild Heerbrugh)

9.3 HORIZONTAL STAFF TACHEOMETRY

9.3.1 The subtense bar

Principles. The horizontal subtense bar is designed for the determination of **horizontal** distances where access is difficult or high accuracy is required. In Fig. 9.13 the required **horizontal** distance D may be calculated from triangle ABC, since

$$D = S/2 \times \cot(\theta/2) \, m$$

where

D is the required distance
S is the fixed distance between subtense bar targets
θ is the parallactic angle measured with the theodolite

Since the distance between the subtense bar targets is fixed at 2 m then $D = \cot(\theta/2)$. This enables the distance to be obtained directly from tables of $\cot(\theta/2)$, once the parallactic angle has been observed.

Accuracy of the system. Since the horizontal plate of the theodolite will measure the horizontal angle even when the bar is above or below the line of collimation, the mean square error of the distance calculated may be expressed as:

$$MD = \pm \frac{D^2}{S} \cdot Ma$$

Tacheometry

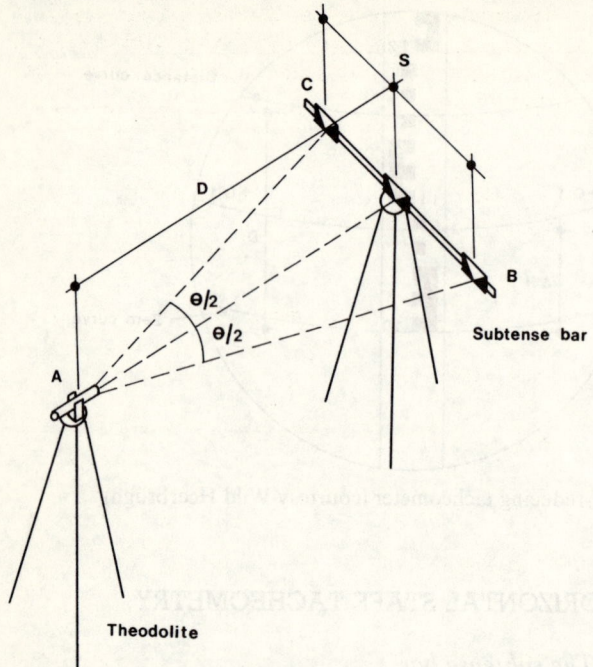

Fig. 9.13 The subtense bar

where

MD = mean square error of the distance
D = distance
S = length of the subtense bar
Ma = angular mean square error

When

S = 2 m and the angular mean square error = 1″

then

$$MD = \pm D^2/2 \times 1''/\theta'' = \Theta^2/400\,000 \text{ mm}$$

This indicates the following:

1. A one-second theodolite should be used with each angle measured by repetition and reiteration serveral times.
2. If the length of the subtense bar were extended the accuracy would increase, although this is impractical. Care is taken during manufacture to reduce any thermal movement problems by means of an invar rod and coil spring.
3. The error increases in proportion to the square of the distance being measured, therefore short distances will give increased accuracy.

Horizontal staff tacheometry

The auxiliary base. In order to reduce the distance being measured and increase the paralactic angle measured the auxiliary base system may be used where the 2 m subtense bar is used to form an auxiliary base as indicated in Fig. 9.14.

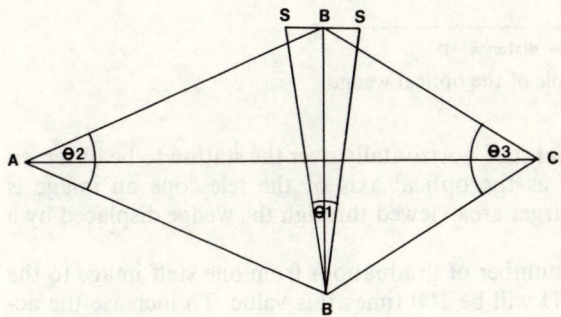

Fig. 9.14 The auxiliary base

In this diagram:

BB = The auxiliary base
SS = 2 m being the subtense bar
AC = the distance to be measured
$\theta_1 = 4° 25' 30''$
$\theta_2 = 6° 15' 40''$
$\theta_3 = 5° 12' 20''$

The length of the auxiliary base = $1 \times \cot \theta_1/2$
$= \cot 2° 12' 45'' = 25.883$ m

$$AD = \frac{25.883}{2} \cot 3° 7' 50'' = 347.874 \text{ m}$$

$$DC = \frac{25.883}{2} \cot 2° 36' 40'' = 283.783 \text{ m}$$

AC = 283.783 + 347.874 = <u>631.657 m</u>

9.3.2 Optical wedge systems

These systems have been developed using the ability of a glass prism to deflect light rays which pass through it by a constant amount (Fig. 9.15). The wedge fits over the object lens of the telescope and covers one third of its area, thus allowing rays which pass through the uncovered area to travel in the normal manner. The rays passing through the prism are deflected by a displacement of distance/100.

Tacheometry

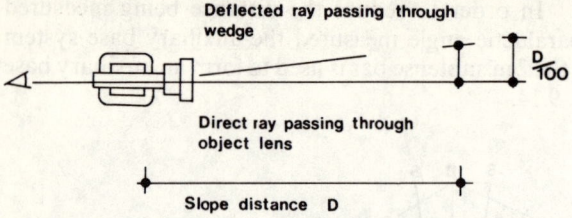

Fig. 9.15 The principle of the optical wedge

By setting a special staff horizontally over the station to be observed at the same height as the optical axis of the telescope an image is observed with the target area viewed through the wedge displaced by a value of D/100.

By counting the number of graduations from one staff image to the other, the distance D will be 100 times this value. To increase the accuracy a vernier may be incorporated above the main scale of the staff and a micrometer screw incorporated on the wedge to displace the line of sight laterally, so that it coincides with a vernier division. The optical wedge equipment is able to provide an accuracy of 1/5000 up to 1/7500 over a distance of 150 m. Care must be taken with the results since it is the **slope distance** between the instrument and horizontal staff which is observed and must be corrected to obtain the horizontal distance.

9.4 EXERCISES

1. (a) In modern surveying practice tacheometry is being applied and used much more than in the past. State under what circumstances tacheometry may be used to replace other methods of surveying, giving the advantages of the use of this method over others.
   (b) The following data was taken with a tacheometer to two staff positions, X and Y, on a drain. The staff was held vertically at each position and the staff readings are in metres.

Staff position	Horizontal circle	Vertical circle	Staff reading		
			Lower	Middle	Upper
A	5° 20′	4° 29′	1.45	1.78	2.11
B	95° 20′	11° 00′	2.15	2.48	2.81

Calculate the difference in level between the two points A and B. The multiplying constant is 100 and there is no additive constant.

(RICS)

Exercises

2. A tacheometer was set up at Station B. The following observations were made in turn, to a staff held vertically at Stations A and C. The staff readings in metres were:

Instrument station	Instrument height	Staff station	Horizontal circle	Vertical circle	Staff reading		
					U	M	L
B	1.82	A	42° 21'	05° 24'	2.84	1.82	0.80
B	1.82	C	173° 46'	08° 14'	2.63	1.82	1.01

The instrument has a multiplying constant of 100' and there is no additive constant. Calculate:
(a) the difference in level between points A and C;
(b) the uniform gradient from A to C. (RICS)

3. A piece of land, roughly rectangular, approximately 150 m by 200 m and varying in level by about 15 m is to be surveyed to produce a plan to a scale of 1:500 showing surface detail and variation in level by contours at 0.5 m intervals. Describe how this work may be undertaken by stadia tacheometry using an automatic vertical index one minute scale reading theodolite and a levelling staff. Include in your description:
(a) the general field procedures assuming four instrument stations and an Ordnance Survey bench mark within sight of one of them;
(b) any formulae to be used in the calculations;
(c) a sketch view of the staff corresponding to the example entry made in (d), below;
(d) a typical page of the fieldbook showing the headings with an entry evaluating horizontal distance and reduced level. (RICS)

10 AERIAL PHOTOGRAPHY

10.1 INTRODUCTION

Photogrammetry is the branch of surveying in which plans, maps and elevations (section 13.4) are prepared by taking measurements from photographs. The procedure has practical advantages where time or access is a critical factor, eg preliminary survey of unaccessible areas or architectural surveys or the upgrading of existing surveys.

10.2 THE PRINCIPLES OF AERIAL PHOTOGRAMMETRY

This is a means of performing a survey and producing a plan by means of aerial photographs. The procedures require **ground control points** to be located by conventional surveying techniques when scale accuracy is important. These procedures may be used for new survey applications or for completing detail or upgrading existing maps where the ground control points have already been located from the original survey, since it will provide the following advantages:

1. Photographs will show the current nature and use of land.
2. Photographs may show many uses, eg soil, forestry, agriculture, geology, etc.
3. The method is useful for initial reconnaissance for roads, railways, etc, since access may not be required.
4. Fieldwork is faster than conventional surveying techniques.
5. The method is most suitable for small-scale work on flat ground when the distortions will be minimal.

The principles of vertical photography

Whilst the procedure will provide the above advantages the equipment is expensive as it tends to be specialised and the considerations discussed in the next section must be made where a true plan is required.

10.3 CONDITIONS REQUIRED FOR A TRUE PLAN

1. The ground to be covered should be horizontal.
2. The camera should be vertical over the detail (max 2°).
3. Atmospheric refraction should be minimal.
4. At high altitudes the earth's curvature must be considered.

10.4 EQUIPMENT

The camera usually incorporates a wide-angle lens giving an angle of view of 75°–95° and a photograph measuring 230 mm × 230 mm with a focal length of 152.4 mm, although super wide angle lenses of up to 120° may be used.

The light rays are considered as being parallel at the lens since the object will be distant from the camera in relation to its focal length (infinite focus). Any oblique rays of light, however, will appear as a dot rather than a point when enlargements of the negative are made.

Some cameras have a series of rings fitted between the lens and the focal plane to improve focusing over a range of distances. This will reduce the 'circles of confusion' caused by an image not being in focus. This phenomenon increases as the camera aperture size increases to allow more light to enter.

An achromatic lens of crown and flint glass is usually incorporated to reduce chromatic aberration of the lens and enable more than one colour of the spectrum to be brought to the same focus.

10.5 THE PRINCIPLES OF VERTICAL PHOTOGRAPHY

Rays from a vertical camera pass through the **perspective centre** of the lens which is identified by **fiducial marks** on the camera lens. The crosses used to identify the principal point may then be referenced to determine any subsequent distortion.

The intersection of the vertical ray on the negative is called the **photograph principal point** (PP) whilst the intersection at ground level is called the **ground principal point** (GP).

Note: In Fig. 10.1:

H is the difference between the aircraft height above sea level (altitude) and the mean height of the terrain.

Aerial photography

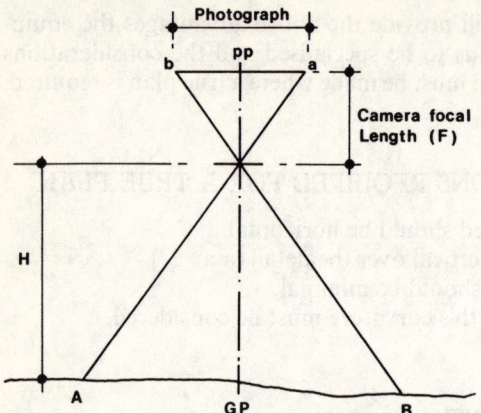

Fig. 10.1 Principles of vertical photography

a–b is the distance measured on the photograph.
A–B is the same distance measured on the ground.

Example 10.1

A camera of focal length f = 152.4 mm produces a photograph of 230 mm × 230 mm incorporating a square of ground 1500 m square. Compute the scale of the photograph and the flying height of the aircraft (Fig. 10.2).

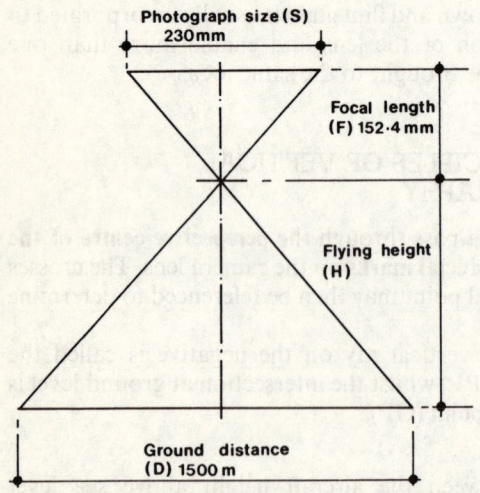

Fig. 10.2 Example of vertical photography

1. Scale of photograph

$$\text{Scale} = \frac{\text{Photograph length}}{\text{Ground distance}}$$

$$= \frac{0.23}{1500}$$

$$= \frac{1}{6522} \quad \text{(Answer)}$$

2. Flying height
This may be determined from similar triangles since:

$$\frac{\text{ground distance}}{\text{photograph size}} = \frac{\text{flying height}}{\text{Focal length}}$$

$$= \frac{1500}{230} = \frac{H}{152.4}$$

Therefore

$$H = \frac{1500}{230} \times 152.4$$

$$= 993.91 \, \text{m}$$

The representative fraction (RF). This is a fraction where the numerator is unity and the denominator is the number of units on the ground, eg

Scale = $1:f/H$
     = $1/H/f$
or   = $1/D/S$ (in Fig. 10.2).

10.6 DISTORTIONS OF AERIAL PHOTOGRAPHS

10.6.1 Distortion due to height relief
In Fig. 10.3 a^1–b^1 represents a negative of AB at ground level with C being a point on a mountain vertically above A.

Photographic distortion. On the negative C is not represented at a^1 although it is vertically above A, but at C' some point distant. This indicates that a photograph is not a plan of the area since changes in

Aerial photography

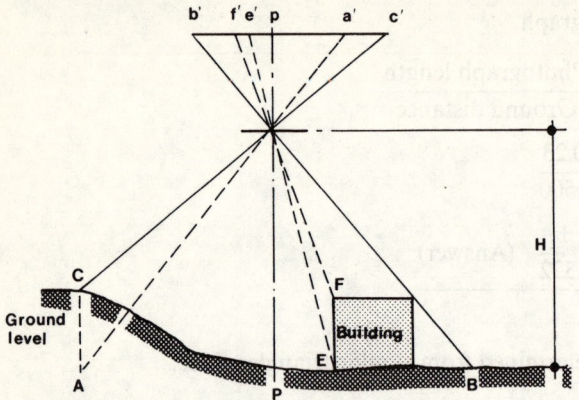

Fig. 10.3 Photographic distortion

altitude will cause a distortion which increases radially from the principal point.

$$\text{The distortion due to height AC} = \frac{AC}{H} \times a'p$$

If points EF are on the side of a building with F vertically above E a plan would only indicate F. On the negative indicated above both the side and roof will be indicated, therefore the building will appear to lean.

Example 10.2

A camera of focal length 152.05 mm was used to photograph a tarn from an aeroplane. If the mean altitude of F − E is 470 m AOD and the horizontal distance is 927 m in Fig. 10.4.

1. Calculate

    (a) the scale of the photograph;
    (b) the altitude of the aircraft.   (RICS)

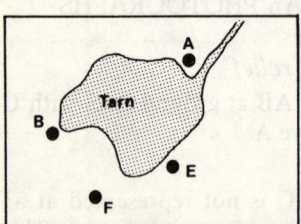

Fig. 10.4 Aerial photograph

Distortions of aerial photographs

(Note: Distance FE measured on the photograph is 82 mm).

(a) Scale = 0.082/927
         = 1/11 305

(b) Height of plane above ground = 0.15205 × 11 305
                                 = 1719 m
                        Altitude = 1719 + 470
                                 = 2189 m

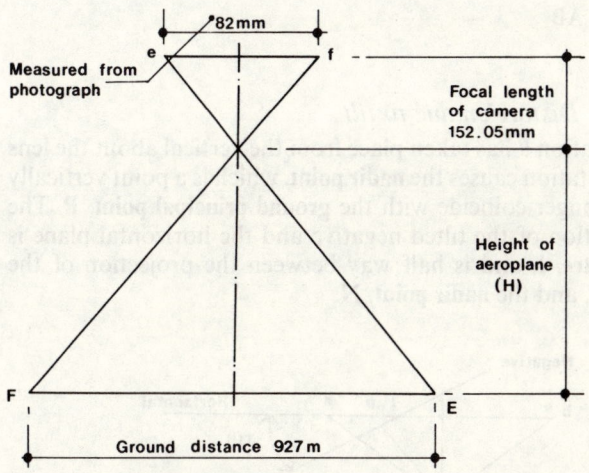

Fig. 10.5 Photograph scale

2. If reduced level of tarn = 375 mm AOD
   What is the scale at this level and length of the tarn?

   Height of plane above AB = 1719 + (470 − 375)
                            = 1814 m

   $$\text{Scale at AB} = \frac{11\,305 \times 1814}{1719} \quad \text{(Fig. 10.6.)}$$

   = 1/11 929

   Line AB = 11 929 × 0.077 (measured from photograph)
           = 918 m   Answer

Note the difference in scale resulting from a change in height relief for a constant altitude as illustrated in Fig. 10.6.

Aerial photography

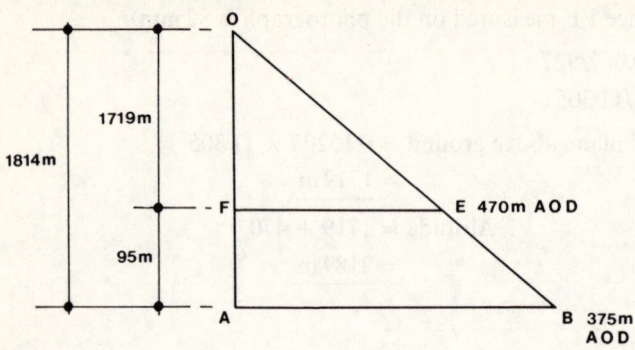

Fig. 10.6 Scale at AB

10.6.2 Distortion due to tilt

In Fig. 10.7 a rotation θ has taken place from the vertical about the lens centre, O. This rotation causes the **nadir point**, which is a point vertically below O to no longer coincide with the **ground principal point**, P. The point of intersection of the tilted negative and the horizontal plane is called the **isocentre**, I, and is half way between the projection of the **principal point**, P, and the **nadir point**, N.

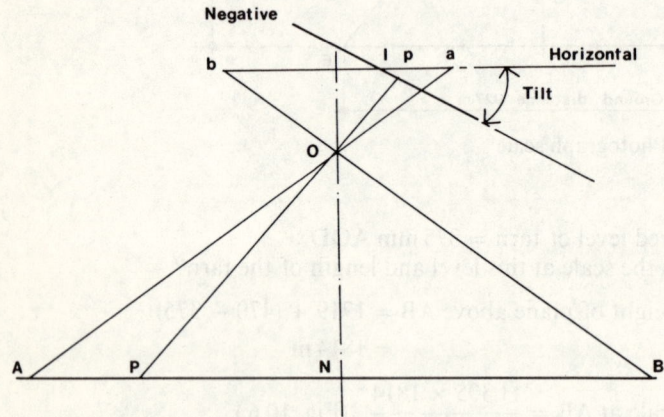

Fig. 10.7 Distortion due to tilt

The distortion caused by this tilt is radial from the **isocentre**, being inward on the same side as the isocentre and outward on the opposite side. To reduce this distortion θ should not exceed 3° or the distortions shown in Fig. 10.8 will result.

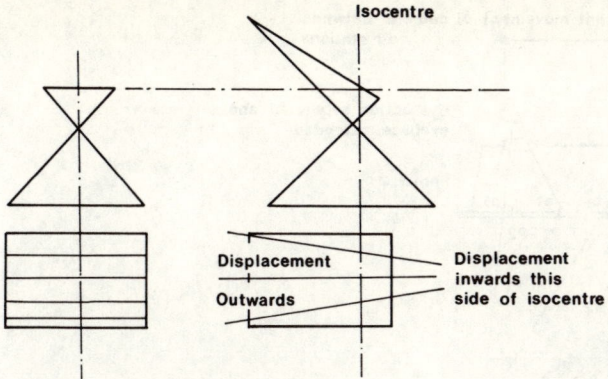

Fig. 10.8 Photographic distortion

10.7 STEREOSCOPY

10.7.1 Introduction

When the human eyes focus on an object the eyes turn inwards until their axes cross at the location of the object. The apparent size of the object can then be determined by the angle subtended by the object at the eye base as illustrated in Fig. 10.9.

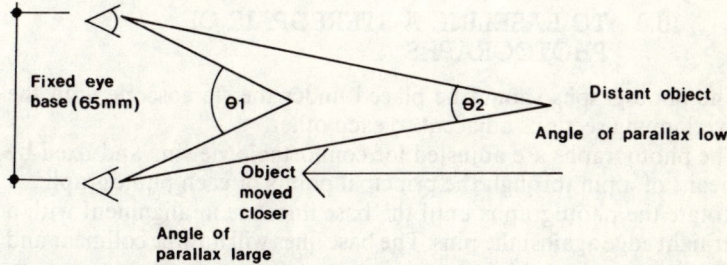

Fig. 10.9 Principle of stereoscopy

The change in the angle of parallax will give a sense of distance to a series of objects.

Whilst a normal eye base of 65 mm will normally be too small to identify changes in ground height on a photograph a stereoscope will increase the distance to 300 mm, which is satisfactory for ground relief perception as illustrated in Fig. 10.10 overleaf.

10.7.2 Stereoscopic pairs

To ensure that each eye sees only one picture a stereoscope is used and a longitudinal overlap of at least 60 per cent is required (50 per cent for stereoscopic examination, 10 per cent margin for tilt).

Aerial photography

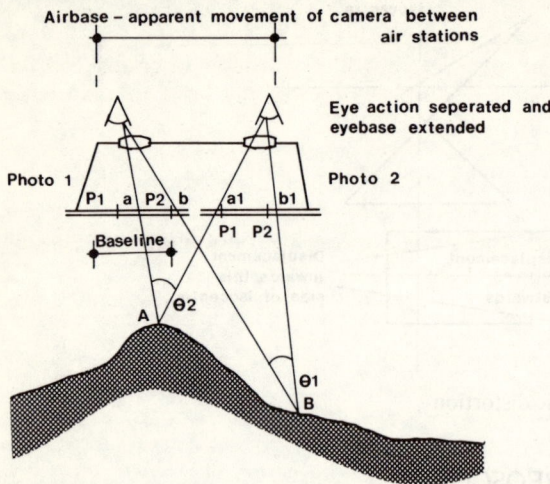

Fig. 10.10 Stereoscopic pairs

Note that there is a common overlap of 20 per cent between the photographs and each principal point is directly below the camera position.

10.8 TO BASELINE A STEREOPAIR OF PHOTOGRAPHS

1. The photographs should be placed under the stereoscope with the overlapping sections adjacent to each other.
2. The photographs are adjusted for comfortable viewing and fixed by means of a pin through the principal points of each photograph.
3. Rotate the photographs until the base lines are in alignment with a straight edge against the pins. The base lines will now be collinear and parallel with the eye base.
4. The photographs may now be taped to the board and the pins removed to form a stereogram.

10.9 RADIAL LINE PLOTTING

This procedure may be adopted where the tilt is limited to 2° and variations in ground height do not exceed 10 per cent. It is often used for revising existing maps.

1. The principal point of each photograph is marked with its own principal point and that of its neighbour, the two points joined with a straight line (**base line**).

Determination of ground height

2. These are then placed under a stereoscope and orientated so that the base lines intersect.
3. Errors due to tilt and relief are controlled by a series of **ground control points**, which have been accurately surveyed on the ground and are clearly visible on both photographs. The base lines and control points are plotted on a transparent sheet which is layed over the photographs to line them in. The ground control points form a rhomboid chain when connected up to the three principal points.

   Minor control points are chosen from the photograph detail and opposite the principal points to build up the strip.
4. A transparent cursor arm marked with a single line is fixed by a pin over each principal point.
5. The cursors are then swung round until they intersect at a detail point, which by means of a linkage system plots the point on a sheet of paper. Detail points must be identifiable on at least two photographs. Since these may then become control points for filling in the topographical detail of the survey.
6. By knowing the exact distance between ground control points the scale may be determined as indicated earlier.

10.10 DETERMINATION OF GROUND HEIGHT

The relative height of features on a stereopair of photographs may be determined by means of a **parallax bar**. The bar consists of a fixed left-hand index glass and a moveable right-hand index glass which can be moved on a circular bar containing a micrometer to record the adjustment made. One revolution of the micrometer head will move the index glass by 0.5 mm along the bar.

By viewing through the stereoscope the fused image of both measuring marks can be presented to the viewer by a slight adjustment of the micrometer. As the micrometer is adjusted this fused image will appear to move vertically above the reference point on the photographs. When the reference point appears to lie on the surface of the photograph then **both** the image and reference mark will have the same **parallax** (see Fig. 10.9) and the micrometer reading is noted. The procedure is then repeated for a further location and the micrometer reading again noted when the fused image appears to lie on the new surface of the photograph, providing the following results:

Micrometer reading of base point P_o = 9.55 mm
Micrometer reading of other point P = 11.05 mm
Flight height of aircraft   H = 3.58 km
Photo base length   f = 89 mm

Aerial photography

The **relative** height between the two locations may be determined as follows:

$$P_d = 9.55 - 11.05 = -1.05 \text{ mm}$$

$$H_d = \frac{-H}{f} \times P_d = \frac{3580}{89} \times 1.5 = -60.30 \text{ mm}$$

ie P is 60.30 m higher than location P_o

Note that the **positive** or **negative** sign in the solution will indicate the position of the location being considered **above** or **below** the reference location.

Where the flying height is not known but at least two **ground control points** have been identified and both the level and location of these are known then the principles outlined in Example 11.2 may be used to determine the flying height.

In Fig. 10.4 the following relationship was identified:

$$\frac{\text{Ground distance}}{\text{Photograph distance}} = \frac{H - h_m}{f}$$

where:

h_m = mean height of the two control points.

H = the flying height of the aircraft.

f = the focal length of the camera.

From this relationship the flying height can be determined with sufficient accuracy for the applications considered in this chapter.

10.11 MOSAICS

These can be rapidly formed from a series of photographs and are useful for general reconnaissance surveys for planners and engineers, since they indicate a wide variety of detail. When the mosaic is referenced to ground control points that have been determined in the field by conventional surveying techniques it is known as a controlled mosaic.

10.12 THE CHARACTERISTICS OF PHOTO IMAGES

(Courtesy the Geographical Association and C. F. Casella and Co Ltd.)
The following characteristics help to identify one image from all other images.

Shape: This is the primary characteristic of any object.

Size: A variety of objects may have similar shapes but few have shape *and* size.

The characteristics of photo images

Tone: The amount of reflected sunlight from a surface will determine the shade of grey on the photo between the extremes of black and white. An object will have a distinguishing tonal value under standard conditions of film and filter, although this may vary depending on the season and time of day.

Texture: This is caused by the frequency in the change of tone. This will depend on the object, the sharpness of the edges and the power of the lens. Texture is also determined by the scale of the photograph; the smaller the scale the more small objects will fuse together.

Shadows: This will reveal features of the side elevation particularly if they fall on level ground. The length of the shadow is controlled by the height of the object, slope of land, time of day and year. A slope facing the sun will shorten the shadow and extend it if it faces away.

$$\text{height} = \text{shadow length} \times \tan \theta$$

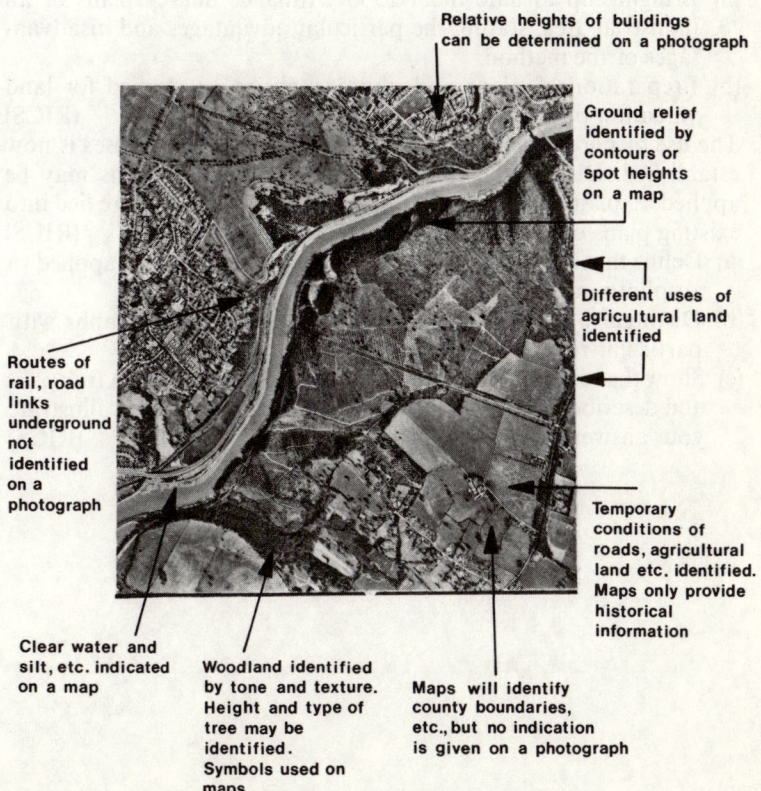

Fig. 10.11 The characteristics of a photograph and map for the interpretation of detail (courtesy C. S. Cassella and Co Ltd)

Aerial photography

Elongated shadows may reveal features that are not normally observed, although in built up areas they tend to obscure shapes.

Pattern: Distinct patterns in a photograph may help to identify an object, eg drainage, orchards, railways, etc.

Distribution: A particular item may be present all over a photograph but its density can vary, eg vegetation on slopes.

A study of the practical applications of aerial photography is given in TRRI report 632, *Inexpensive Aerial Photography for Highway Engineering and Traffic Studies*, published by the Transport and Road Research Laboratory (Department of the Environment).

10.13 EXERCISES

1. Discuss the use of aerial surveying for:
   (a) Bringing up to date the 1/2500 Ordnance Survey plans of an industrial area, stating the particular advantages and disadvantages of the method.
   (b) Preparation of plans and photographs to be studied for land valuation purposes. (RICS)
2. The use of photography from the air for surveying purposes is now established. Discuss the uses to which such photographs may be applied. Explain how such aerial surveys or photographs are tied into existing plans or information. (RICS)
3. (a) Define the terms principal point, nadir and isocentre as applied to air photographs.
   (b) Distinguish between vertical and oblique air photographs with particular reference to scale.
   (c) Show that air photographs cannot normally be used as a true plan and describe the two geometric distortions responsible. Illustrate your answers with clear sketches. (RICS)

11 AREAS AND VOLUMES

11.1 INTRODUCTION TO AREAS

The surveyor is often required to calculate the area of a plot of land from a range of maps, plans, working drawings, etc. The accuracy of such calculations will be determined by the application being considered, the data available and the method of calculation used.

Where a small area is being determined from a large-scale map or plan, the value is often expressed in m², whilst for most practical applications it is more convenient to use the hectare, where:

1 hectare = $10\,000\,\text{m}^2$

11.2 AREA CALCULATION PROCEDURES

The procedure adopted will depend on the accuracy required, the shape of the survey and the assumptions made when undertaking the calculations. The shape of the figure will be either:

(a) defined by straight lines;
(b) an irregular boundary on one or more sides.

11.2.1 An area defined by straight lines

Where a survey can be divided into a number of straight sided figures that can be related to basic mensuration formulae, the lengths of each side may be obtained from fieldnotes or plans and the following formulae used, as illustrated in Fig. 11.1.

Area of a triangle.

$\text{Area} = \tfrac{1}{2} b \times h$
$\text{Area} = \sqrt{s(s-a)(s-b)(s-c)}$ where $s = \tfrac{1}{2}(a+b+c)$
$\text{Area} = \tfrac{1}{2} bc \cdot \sin\theta$

Area calculation

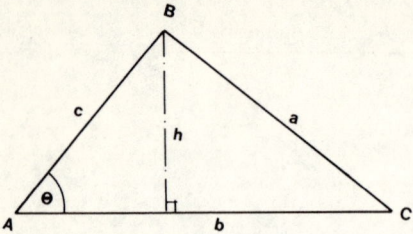

Fig. 11.1 The triangle

In Fig. 11.2 it can be seen that: Area of trapezium = $\frac{1}{2}$(a + b)h where a, b are the lengths of the parallel sides, h is the distance between them.

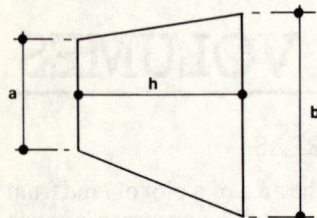

Fig. 11.2 The trapezium

Area of a circle = πr^2 (where r is the radius)

The sum of the areas of each individual figure will then provide the total **plan area** of the survey. When the area under consideration slopes at a **constant** gradient the **true area** is given by:

True area = plan area × secant of slope

If the ground slopes in more than one direction each gradient must be considered individually and the sum of each unit will then provide the total true area.

11.2.2 Irregular figures

Many plans are not defined by regular geometric shapes and either **graphical** methods or **calculation** procedures may be found beneficial. The graphical methods possible are as follows:

Squares. Squared tracing paper may be placed over the plotted survey and all the covered squares counted. The size of the grid should be selected with regard to:

(a) the accuracy required, since small squares will increase the accuracy but also lengthen the time taken;
(b) the scale of the survey;
(c) ease of calculation from the area represented by one square.

Area calculation procedures

All the full squares are counted and where a square is **not completely filled** the area may be determined by:

1. Scale off the remaining area into triangles or other rectangular forms capable of being measured by standard mensuration formulae. Detail 1 in Fig. 11.3 indicates the boundary as a diagonal leaving triangle ABC to be added to the complete squares.

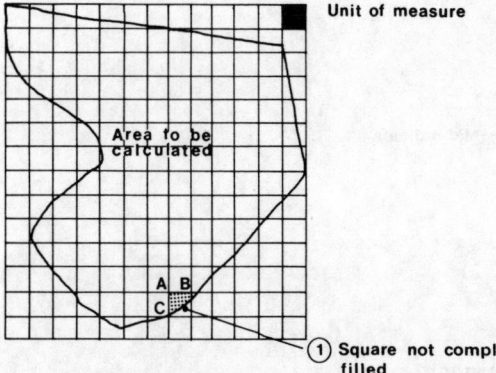

Fig. 11.3 The square grid

2. Judge fractional parts by eye. The line is drawn so that the area of land that has been excluded is equal to the space that has not been filled, ie Area B = Area A. The total area is assumed to be equal to the area of triangle GEH in Fig. 11.4.

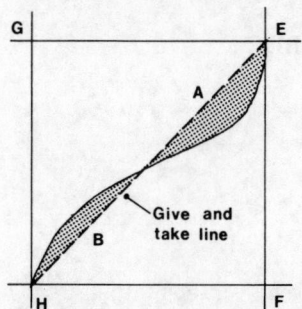

Fig. 11.4 Give and take lines

Mid-ordinate rule. If the tracing paper is ruled in parallel strips instead of squares, the width being some number of whole units, then the same principle as the grid may be applied. The procedure will be reduced if a mid-ordinate is drawn down each strip and the area determined by

Area calculation

multiplying the length of the strip by the width and then adding them all together as illustrated in Fig. 11.5.

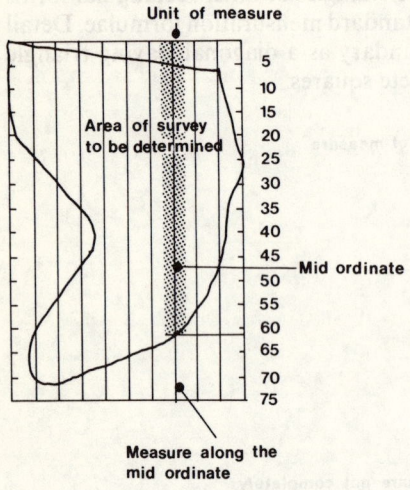

Fig. 11.5 The mid-ordinate graph

11.2.3 Calculation procedures

Calculation of the rectilinear area. Where a chain survey has been carried out, the area enclosed by the main chain lines will be defined by a series of triangles which constitute the **rectilinear area**. By applying the triangle formulae in section 11.2.1 the area may be determined as follows. In Fig. 11.6, area enclosed by ABCD may be calculated and the calculation checked since:

$$\text{Area ABCD} = \text{Area ABD} + \text{Area BDC}$$

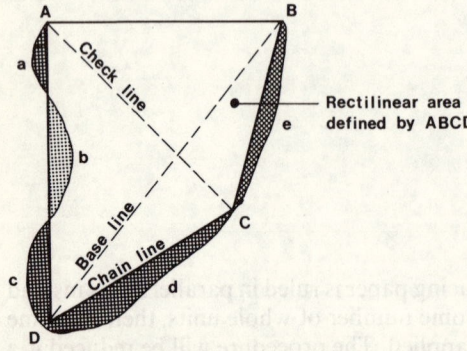

Fig. 11.6 Survey example

Area calculation procedures

This may be calculated since the lengths of the sides are known and

$$\text{Area ABD} = \sqrt{s(s-a)(s-b)(s-d)} \left(\text{where: } S = \frac{(a+b+c)}{2} \right)$$

Also the above may be checked since:

Area ABCD = area ACB + area ACD

Correction to rectilinear area. The true area of the survey outlined in Fig. 11.6 may be determined from the rectilinear area since areas, A, C, D and E must be added, whilst area B will be deducted.

Since the perimeter of the survey has been determined by means of offsets from the traverse legs one of the following procedures may be used:

The trapezoidal rule. Since each offset is taken at 10 m intervals the area may be determined by use of the trapezium formulae in section 11.2.1 for each pair of offsets as illustrated in Fig. 11.7:

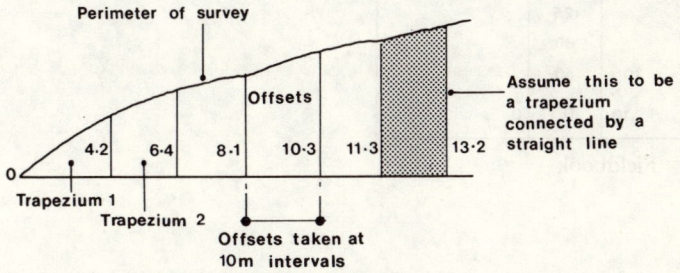

Fig. 11.7 Application of the trapezoidal rule

$$\text{Area} = 10 \left[\frac{0 + 4.2}{2} + \frac{4.2 + 6.4}{2} + \frac{6.4 + 8.1}{2} + \frac{8.1 + 10.3}{2} \right.$$
$$\left. + \frac{10.3 + 11.3}{2} + \frac{11.3 + 13.2}{2} \right]$$

Since each offset value appears twice except the first and last the formulae may be reduced to:

$$\text{Area} = 10 \left[\frac{0 + 13.2}{2} + 4.2 + 6.4 + 8.1 + 10.3 + 11.3 \right]$$

- Width between strips
- × Half sum of first and last ordinate
- + Sum of remaining ordinates

Area calculation

The trapezoidal rule may be used where short offsets have been taken at regular intervals or where the lengths are scaled from a plan and states:

The area is equal to half the sum of the first and last ordinate plus the sum of the remaining ordinates multiplied by the offset interval.

The above formulae may also be applied to a chain line running through a narrow strip of land with offsets on either side as indicated below.

Example 11.1

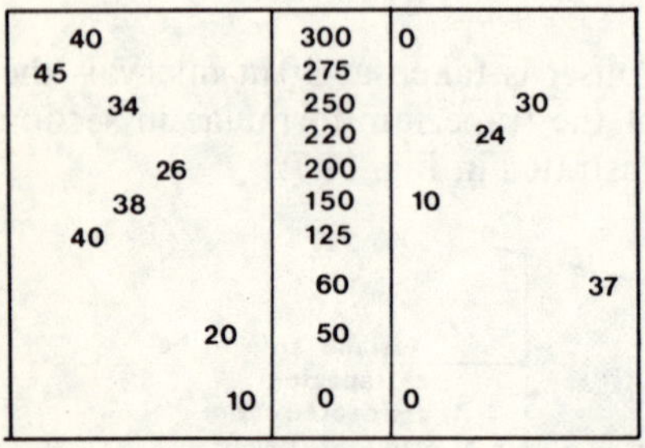

40		300	0
45		275	
	34	250	30
		220	24
	26	200	
	38	150	10
40		125	
		60	37
	20	50	
	10	0	0

Fig. 11.8 Fieldbook

Compute the area enclosed by the survey data shown in Fig. 11.8. Procedure is as follows.

1. Identify uniform intervals where the trapezoidal rule may be used.
2. Identify any individual trapeziums.
3. Identify any other rectilinear figures.
4. Any remaining areas must be determined by approximation.
5. Add all the above areas together to determine the total area.

In the example above:

Left-hand side
Trapezoidal rule (25 m intervals)

$$25\left[\frac{40 + 34}{2} + 45\right] = 2050 \text{ m}^2$$

Trapezoidal rule (50 m intervals)

$$50\left[\frac{34 + 38}{2} + 26\right] = 3100 \text{ m}^2$$

Right-hand side
(no uniform intervals)
Area of first and last triangle

$$\frac{50}{2} \times 30 = 750 \text{ m}^2$$

$$\frac{60}{2} \times 37 = 1110 \text{ m}^2$$

Area calculation procedures

Left-hand side
Area of remaining trapeziums
(interval not uniform)

$$\frac{38 + 40}{2} \times 25 = 975 \, m^2$$

$$\frac{40 + 20}{2} \times 75 = 2250 \, m^2$$

$$\frac{20 + 10}{2} \times 50 = 750 \, m^2$$

Total area of
left-hand side = $\underline{9125 \, m^2}$

Right-hand side
Area of remaining trapeziums

$$\frac{30 + 24}{2} \times 30 = 810 \, m^2$$

$$\frac{24 + 10}{2} \times 70 = 1190 \, m^2$$

$$\frac{10 + 37}{2} \times 90 = 2115 \, m^2$$

Total area of
right-hand side = $\underline{5975 \, m^2}$

Total area = Area on left-hand side + area on right-hand side.
= $9125 \, m^2 + 5975 \, m^2$
= $\underline{1.510 \, hectares}$

The accuracy of the trapezoidal rule may be improved by reducing the uniform interval between ordinates although the length of the calculation is increased. One advantage of the rule is that it may be used for any number of ordinates providing the interval remains uniform.

Simpson's rule. The above rule assumes that each pair of ordinates is connected by a straight line but this is only true for narrow strips. Simpson's rule assumes the intervals are connected by a parabola as indicated in Fig. 11.9, although uniform intervals are still required:

Area enclosed by parabola = $2/3 \, (p \times 2d)$

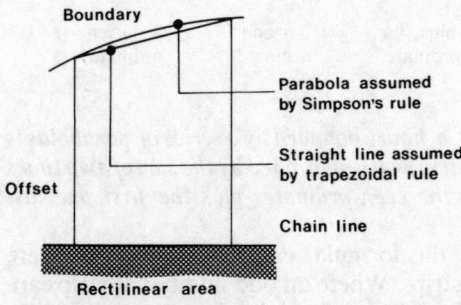

Fig. 11.9 The parabola

Area calculation

where:

P = mid-ordinate
d = distance between mean of end ordinates

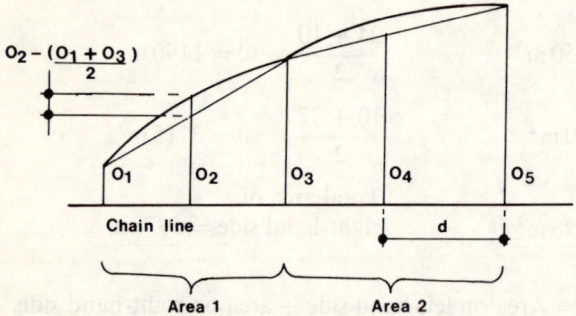

Fig. 11.10 The application of Simpson's rule

Area 1 = Area of trapezoid + area defined by the parabola

$$= \left(\frac{0_1 + 0_3}{2}\right) \times 2d + \frac{2}{3}\left[0_2 - \left(\frac{0_1 + 0_3}{2}\right)\right] \times d$$

$$= \frac{d}{3}\left[0_1 + 0_3 + 40_2\right]$$

Similarly:

$$\text{Area 2} = \frac{d}{3}\left[0_3 + 0_5 + 40_4\right]$$

$$\text{Total area} = \frac{d}{3}\left[(0_1 + 0_5) + 20_3 + 4(0_4 + 0_2)\right]$$

- interval
- first plus last ordinate
- 2 × odd ordinate
- 4 × even ordinates

The rule states: *The area of a figure bounded by a series of parabolas is equal to one third the uniform strip width multiplied by the sum of two times the odd ordinates, four times the even ordinates plus the first and last ordinate.*

As indicated by the proof, this formulae can only be applied where there are an **even** number of strips. Where an odd number of strips are identified the last strip must be calculated separately.

The planimeter

To identify the ordinates it is useful to number each offset starting with ordinate 1 as indicated in the proof.

Example 11.2
Calculate the area in Fig. 11.11 by Simpson's rule.

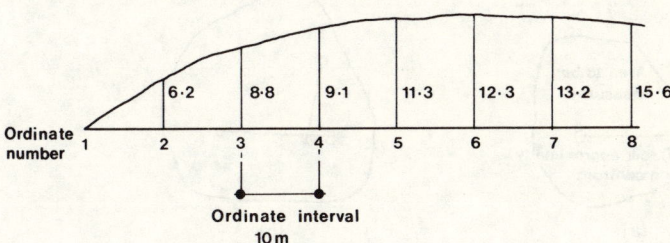

Fig. 11.11 Area example

1. Number each ordinate.
2. Note that there are an even number of ordinates.
3. Use Simpson's rule for ordinates 1–7. Calculate the remaining area as a separate trapezium.

Simpson's rule for ordinates 1–7 gives the following.

$$\text{Area} = \frac{10}{3}\left[\frac{0 + 13.2}{2} + 2(8.8 + 11.3) + 4(6.2 + 9.1 + 12.3)\right] \text{m}^2$$

$$= \frac{10}{3}[6.6 + (2 \times 20.1) + (4 \times 27.6)]\,\text{m}^2$$

$$= 3.3(6.6 + 40.2 + 110.4)\,\text{m}^2 = 518.76\,\text{m}^2$$

Area of remaining trapezium $= 10\left[\dfrac{13.2 + 15.6}{2}\right] = 10 \times 28.8 = 288\,\text{m}^2$

Total area $= 518.76 + 288 = 806.8\,\text{m}^2 = \underline{0.08\text{ hectares}}$

11.3 THE PLANIMETER

This instrument offers the simplest and quickest method of measuring the area of an irregular shaped figure and consists of the following:

The tracer arm. This is attached to the measuring unit and supports a tracer point for tracing the perimeter of the area to be calculated.

The pole arm. This pivots on a weighted pole block which remains stationary during operation.

Area calculation

The measuring unit. The pole and tracer arm above are linked by a measuring unit which consists of a revolving wheel in contact with the paper and a revolution counter containing a vernier.

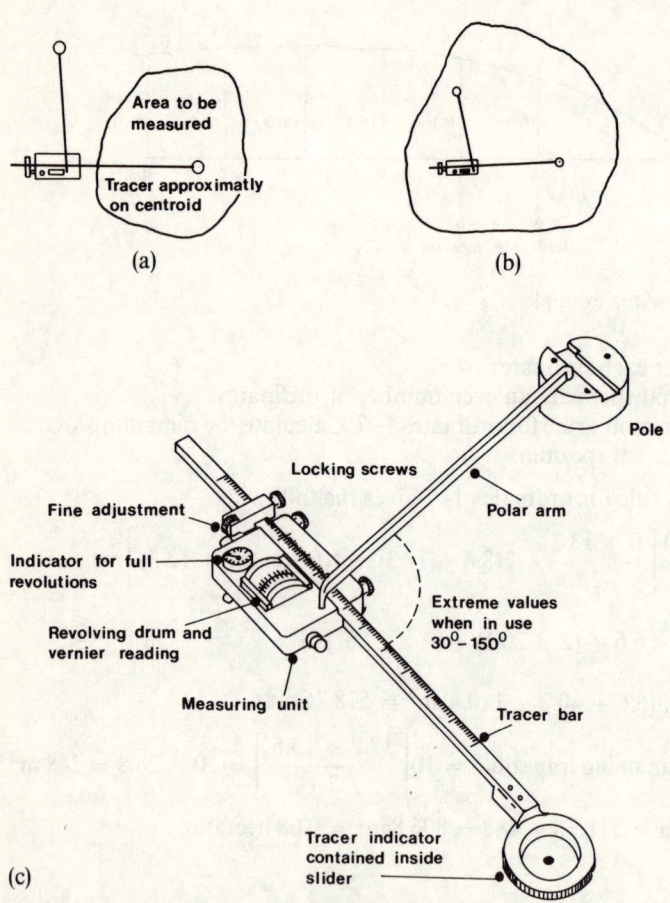

Fig. 11.12 Application of the planimeter (a) pole outside area (b) pole inside area to be measured (c) the polar planimeter (courtesy Carl Zeiss Zena Ltd)

The instrument determines the area by recording the number of revolutions made by the circular drum when the tracer point is guided over the perimeter of the area to be determined. A constant is provided with the instrument to indicate the area represented by one revolution of this measuring wheel.

Area = Number of revolutions recorded × area per revolution factor

Calculating areas from rectangular co-ordinates

In some instruments the length of the tracer arm may be adjusted and a set of scales are provided to set the tracer arm at a predetermined position depending on the scale of the survey.

The planimeter may be used with the pole weight either inside or outside the area to be determined as indicated in Fig. 11.12(a) and (b).

Pole outside area to be determined. This procedure is adopted where possible since it is a less complex procedure, providing the whole area can be traced. If it is not possible to cover the whole area in this position, the survey may be sub-divided into a number of small areas, each determined as follows:

1. Choose a point on the perimeter where the measuring wheel does not revolve for a small movement of the tracer point.
2. Note the reading on the measuring unit or, where a zero setting plunger is incorporated, set the display to zero.
3. Carefully trace the perimeter of the survey **in a clockwise** direction, returning to the starting point.
4. Note the new reading of the measuring unit.
5. The difference in reading between observations 2 and 4 is proportional to the area traced.
6. Repeat the above procedure until **three** consistent readings are observed.
7. Multiply the mean reading by the appropriate area per revolution factor to obtain the area bounded by the perimeter traced.

Pole point inside the area to be traced. Where the instrument is set up as in Fig. 11.12(b) it is possible to trace a whole circle without the measuring wheel moving. This circle is called the zero circle and a constant is provided with the instrument called the zero circle constant. This constant indicates the area traced by the whole zero circle movement.

If the area to be traced is greater than the zero circle then this constant must be added to the value obtained. If the reading is less than the zero circle constant it is subtracted. The result obtained is then multiplied by the area per revolution factor.

11.4 CALCULATING AREAS FROM RECTANGULAR CO-ORDINATES

If the rectangular co-ordinates of the survey stations are known, then the **rectilinear area** of the survey may be determined **without** having to plot the survey. The area enclosed by the survey ABCD illustrated in Fig. 11.13 may be determined from the sum of the areas of the following **trapeziums**:

$$\text{Area ABCD} = \text{ABYX} + \text{BCZY} - \text{ADWX} - \text{BCZY}$$

Area calculation

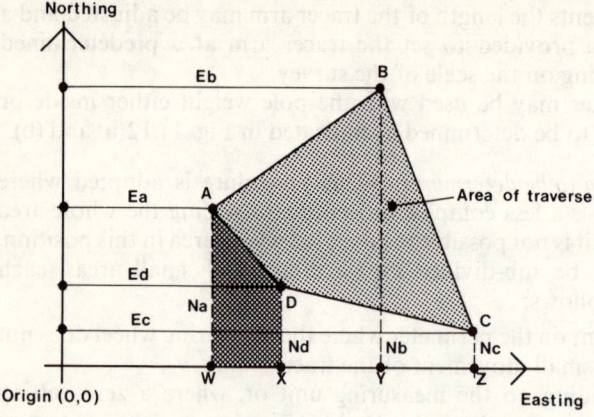
Fig. 11.13 The traverse survey

Substituting the **easting** (E) and the **northing** (N) of each station into the above equation will provide;

$$\text{Area ABCD} = \left[\frac{(N_a + N_b)}{2}(E_b - E_a)\right] + \left[\frac{(N_b + N_c)}{2}(E_c - E_b)\right]$$

$$- \underbrace{\left[\frac{(N_a + N_b)}{2}(E_d - E_a)\right]}_{\text{(area ADWX)}} - \underbrace{\left[\frac{(N_d + N_c)}{2}(E_c - E_d)\right]}_{\text{(area BCZY)}}$$

The above expression will reduce to:

Area ABCD
$$= \tfrac{1}{2}[E_a N_d - E_a N_c + E_b N_a - E_b N_c + E_c N_b - E_c N_d + E_d N_c - E_d N_a]$$
$$= \tfrac{1}{2}[E_a(N_d - N_c) + E_b(N_a - N_c) + E_c(N_b - N_d) + E_d(N_c - N_a)]$$

Easting of station A ↑
Northing of preceding station ↑
Northing of following station ↑

In each case the **easting** of each station is multiplied by the difference between the **northing** of the preceding and the following stations, therefore the following general rule may be stated:

The area enclosed by the legs of a traverse (rectilinear area) is given by half the sum of the product of the easting of each station multiplied by the product of the northing of the preceding station minus the northing of the following station.

Example 11.2

Determine the area of the square tranverse illustrated in Fig. 11.14 where the length of each side is 50 m.

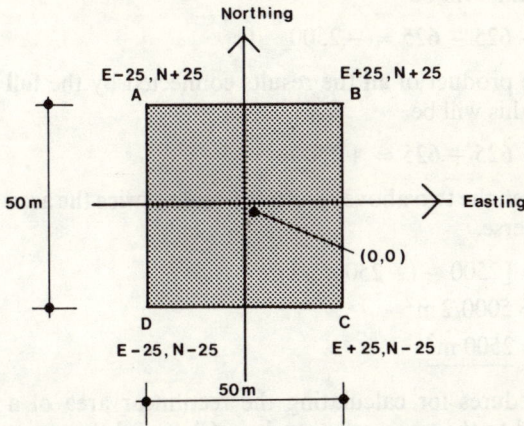

Fig. 11.14 Traverse example

$$\text{Area ABCD} = \tfrac{1}{2}[-25(-25-25) + 25(25+25)$$
$$+ 25(25+25) - 25(-25-25)]\,\text{m}^2$$
$$= \tfrac{1}{2}[(-25 \times -25) - (25 \times 50) - (25 \times 50) - (25 \times -50)]\,\text{m}^2$$
$$= \tfrac{1}{2}(1250 + 1250 + 1250 + 1250)\,\text{m}^2$$
$$= 5000/2\,\text{m}^2$$
$$= \underline{2500\,\text{m}^2}$$

Area of square having sides 50 m long is 2500 m² from basic mensuration formulae, therefore formulae is correct.

Note from the above example that care must be taken to identify the **positive** or **negative** sign of each **latitude** and **departure** since these must be correctly accounted for in the equation.

When the rectangular co-ordinates of each station have been recorded in the format provided by the **traverse table**, with the first

Table 11.1 Rectangular co-ordinates

Station	Easting (m)	Northing (m)
A	−25	25
B	25	25
C	25	−25
D	−25	−25
A	25	25

Area calculation

station repeated at the end of the table, then the following procedure may be followed:

1. Find the sum of the product of **all** the results connected by the **dotted** lines. In Table 11.1 this will be

    $-625 - 625 - 625 - 625 = -2500$

2. Find the sum of the product of **all** the results connected by the **full** lines. In Table 11.1 this will be

    $+625 + 625 + 625 + 625 = +2500$

3. The **difference** between the two above results is equal to **twice** the area enclosed by the traverse.

    Area ABCD $= [2500 - (-2500)]/2\,\text{m}^2$
    $= 5000/2\,\text{m}^2$
    $= 2500\,\text{m}^2$

Both of these procedures for calculating the rectilinear area of a traverse may be applied to the traverse example in Ch. 6 and the result verified by inspection of the computer print-out illustrated in Table 6.7. The program used to produce the print-out includes the above calculations.

11.5 VOLUME CALCULATIONS

The calculation of the volume of earthworks is an important prerequisite for many construction projects since it enables the contractor to:

(a) determine the plant capacity required to undertake the proposed development;
(b) estimate the time to be taken for the excavation or fill work;
(c) determine a method of disposal for the excavated material;
(d) determine the quantity of fill material to be obtained.

The calculation procedure selected will be influenced by the following:

(a) The information available.
(b) The profile of the excavation or fill.
(c) The accuracy desired.

The results are expressed in **cubic metres** (m^3) and can only be approximated due to the problems of bulking when the excavated material expands or on compaction of fill material.

The basic formula for a prismoid having parallel cross-sections is:

Volume = cross-sectional area
× distance between the selected cross-sections

The above formula relies on straight lines connecting the profiles selected but this may not be true in practice. This error will be reduced if the cross-sections are carefully selected and the distance between is reduced.

11.6 CROSS-SECTIONS

In order to apply the basic mensuration formulae each cross-sectional area must be estimated within the accuracy of the information available. In order to improve the accuracy when estimating the cross-sectional area from plotted sections the **vertical** scale is often exaggerated. The cross-sectional area of a proposed section may be determined by;

Measuring the plotted cross-sections. This may be undertaken graphically as described in section 11.2.2 or by use of the **planimeter** (section 12.3 and then corrected for any scale differences as follows:

Example 11.3

The area of a particular cross-section detailed by a survey is 19.27 cm^2 when measured with a planimeter. What area does this represent if the vertical scale is 1:50 and the horizontal scale is 1:150?

$$\text{True area} = \frac{19.27 \times 150 \times 50}{100 \times 100}$$

$$= 14.45 \, \text{m}^2$$

Area calculation from basic principles. Basic mensuration formulae may be used where the cross-section is composed of triangles, rectangles or a trapezium. Where an embankment or cutting is on sloping ground as illustrated in Fig. 11.15 such that:

Height at centre $= h$
Formation width $= b$
Side slope $= S:1$ (1 vertical)
Ground slope $= g:1$ (1 vertical)

Then:

$$W_1 = \frac{g(b/2 + hs)}{g - s}$$

$$W_2 = \frac{g(b/2 + hs)}{g + s}$$

$$\text{Area} = \frac{W_1 W_2 - (b/2)^2}{s}$$

Area calculation

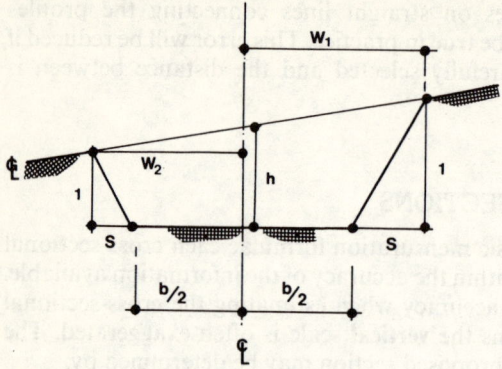

Fig. 11.15 Ground slope

11.7 VOLUME BY MEAN AREAS

After calculation of the end areas of a proposed excavation the volume may be estimated by application of the basic formulae for volumes, as follows:

$$\text{Volume} = \frac{d(A_1 + A_2)}{2}$$

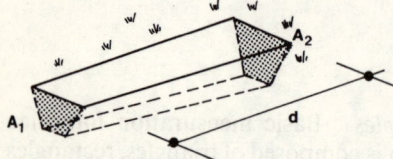

Fig. 11.16 End areas

Providing the area of the mid-section is the **mean** of the end cross-sectional areas the above expression may be applied. Where several cross-sections are placed an equal distance apart as indicated in Fig. 11.17 then,

$$\text{Volume} = d\left(\frac{A_1 + A_2}{2}\right) + d\left(\frac{A_2 + A_3}{2}\right) + d\left(\frac{A_3 + A_4}{2}\right)$$
$$+ d\left(\frac{A_{n-1} + A_n}{2}\right)$$

$$= \frac{d}{2}(A + 2A + 2A + 2A + 2A + A_n)$$

$$= \frac{d}{2}(\text{First area} + \text{last area} + 2 \times \text{remaining areas})$$

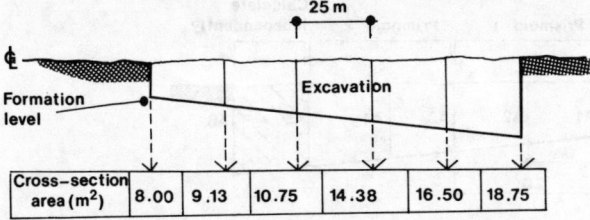

Fig. 11.17 Excavation example

The above formulae is often called the **trapezoidal rule for volumes** and may be applied to any number of cross-sections providing they are all **equally** spaced apart.

Example 11.7

$$\text{Volume} = \tfrac{25}{2}((8.00 + 18.75) + 2(9.13 + 10.75 + 14.38 + 16.50))$$
$$= \tfrac{25}{2}(26.75 + 101.52)$$
$$= \underline{1603.38 \text{ m}^3}$$

11.8 PRISMOIDAL FORMULAE

Where the volume between parallel cross-sections is such that it may be considered as a **prismoid** since each cross-section is **not** connected by **straight** lines, a more accurate estimate of the volume can be made since,

The volume of a prismoid is given by;

$$\text{Volume (V)} = \frac{d}{6}(A_1 + 4M + A_2)$$

Where

A_1, A_2 are the end areas

M is the **calculated** area of the mid-section (it is not assumed to be the MEAN of the end areas)

d is the distance between the end areas

By treating long excavations or proposed fill work as a series of prismoidal forms, as illustrated in Fig. 11.18, with alternate areas a distance 2D apart then,

$$V_1 = \frac{2D}{6}(A_1 + 4A_2 + A_3)$$

$$V_2 = \frac{2D}{6}(A_3 + 4A_4 + A_5)$$

Area calculation

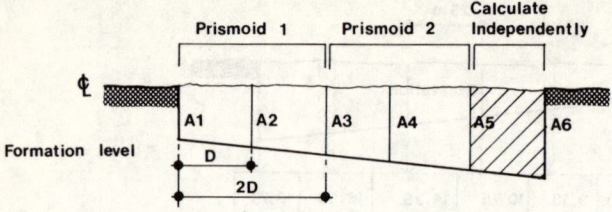

Fig. 11.18 Prismoidal rule

Since $V_{tot} = V_1 + V_2$

$$V_{tot} = \frac{2D}{6}(A_1 + 4A_2 + A_3 + A_3 + 4A_4 + A_5)$$

$$= \frac{D}{3}(A_1 + 4(A_2 + A_4) + 2A_3 + A_5)$$

This expression for volumes is called **Simpson's rule** for volumes and states that the volume is:

The sum of the areas of the end cross-sections plus four times the area of the even cross-sections plus twice the sum of the odd cross-sections multiplied by one third the distance between cross-sections.

The rule requires an **odd** number of ordinates for its application and where an even number of cross-sections have been identified the last prismoid must be calculated separately. This procedure is illustrated in the example below;

Example 11.5

The cross-sectional areas of the excavation in Fig. 11.18 are as follows;

Cross-section	A_1	A_2	A_3	A_4	A_5	A_6
Area (m²)	11.21	15.13	17.90	19.23	18.31	19.65

If each cross-section is 15 m apart, calculate the volume of the excavation.

Volume between cross-sections $A_1 - A_5$

$V_{1-5} = \frac{15}{3}((11.21 + 18.31) + 4(15.13 + 19.23) + 2(17.90))\,m^2$
$= \frac{15}{3}(29.52 + 137.44 + 35.80)\,m^2$
$= \underline{1013.80\,m^3}$

$$V_{5-6} = \tfrac{15}{2}(18.31 + 19.65)\,\text{m}^3$$
$$= 284.70\,\text{m}^3$$

Total vol = 1013.80 + 284.70
        = 1298.50 m³

11.9 CROSS-SECTIONS COMBINING CUT AND FILL SECTIONS

Both the method of end areas and the prismoidal rule may be used for both cuttings and fill applications, adopting the procedures illustrated in the last example. When the cut or fill does not follow the general ground slope the sections selected at uniform intervals may include **both** cut and fill details.

If all the sections include both cut and fill as illustrated in Fig. 11.19 the end area or prismoidal rule may be applied to the cut on each section and then the fill for each section.

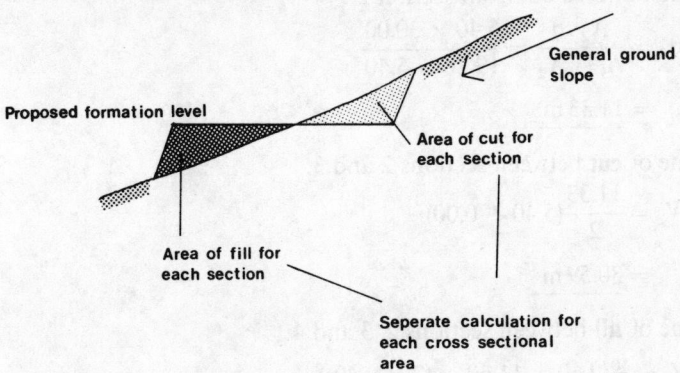

Fig. 11.19 Cut and fill sections

When the earthwork commences in cut then terminates with fill sections and a uniform gradient between sections may be assumed, the position of **zero** cut and fill must be identified as follows;

Assume three cross-sections a horizontal distance d m apart have the following properties;

A_1 represents an area of cut.
A_2 represents an area of cut and fill.
A_3 represents an area of fill.

Assuming the location where the cut is **zero** is from A_2–d_0. Then

$$\frac{A_2 - 0}{d_0} = \frac{A_1 - A_2}{d} \quad \therefore \quad d_0 = \frac{A_2 - d}{A_1 - A_2}$$

Area calculation

Example 11.6

Given the following information regarding a proposed excavation, calculate the quantity of fill to be imported, assuming the cut material may be used as fill. Each cross-section has been taken at 30 m intervals.

	Cut (m^2)	Fill (m^2)
Cross-section 1	19.70	0.00
Cross-section 2	5.40	2.60
Cross-section 3	0.00	9.40
Cross-section 4	0.00	11.60

1. Volume of **cut** between sections 1 and 2 may be calculated from the end areas:

$$V_c = \tfrac{30}{2}(19.70 + 5.40)$$
$$= \underline{376.50 \text{ m}^2}$$

2. Location of **zero** cut from section 2:

$$d_0 = \frac{A_2 \cdot d}{A_1 - A_2} = \frac{5.40 \times 30.00}{19.70 - 5.40}$$
$$= \underline{11.33 \text{ m}}$$

3. Volume of **cut** between sections 2 and 3:

$$V_c = \frac{11.33}{2}(5.40 + 0.00)$$
$$= \underline{30.59 \text{ m}^3}$$

4. Volume of **fill** between sections 2, 3 and 4:

$$V_f = \tfrac{30}{2}(2.60 + 11.60 + (2 \times 9.40))$$
$$= \underline{495 \text{ m}^3}$$

5. Location of **zero** fill from section 2:

$$d_f = \frac{2.60 \times 30}{9.40 - 2.60}$$
$$= \underline{11.47 \text{ m}}$$

6. Volume of **fill** between section 2 to location of zero fill:

$$V_f = \frac{11.47}{2}(2.60 + 0.00)$$
$$= \underline{29.82 \text{ m}^3}$$

7. Total **fill** to be transported:

   Total amount of cut = 30.59 + 376.50 = 407.09 m³
   Total amount of fill = 29.82 + 495.00 = 524.82 m³
   Total fill to be imported   117.73 m³

   Total fill to be imported is 117.73 m³

11.10 VOLUMES FROM CONTOURS

If the survey of an area includes contour information and a uniform slope between each contour may be assumed, then the volume of material between each contour can be determined.

This is possible since the respective area enclosed by each contour may be determined by use of the **planimeter** (section 11.3) and the results incorporated into the **prismoidal** formulae. By treating alternate areas as mid-areas in the formulae or interpolating between each cross-section and using the contour interval as the distance between each cross-section the volume may be determined. This procedure will not be reliable where the ground slope between each contour is not uniform and the contour interval is greater than 2 m, and these restrictions must be considered when undertaking the survey.

Since the prismoidal formulae requires an **even** number of intervals, any application which includes an odd number of contour intervals will require the following adjustment:

(a) determine the last volume on top of a hill as a separate calculation;
(b) determine the last volume at the bottom of a valley as a separate calculation.

The volume obtained from the above adjustment will be small and is added to the result of the prismoidal formulae. Alternatively the trapezoidal rule for volumes may be used without any need for the above corrections.

Example 11.7

The following results were provided by a survey of a proposed fill site.

   Contour (m)   50   48   46   44   42
   Area enclosed (m) 202 188 164 150 120

Calculate the fill capacity available.

Volume of fill = $\frac{4}{6}$[202 + 4(188 + 150) + 2(164) + 120]
              = $\frac{2}{3}$[202 + 1352 + 328 + 120]
              = 1335 m³

11.11 VOLUMES FROM SPOT HEIGHT

This procedure may be followed providing the surveyor has divided the survey area into **squares, rectangles** or **triangles** and has recorded the **spot level** of each grid intersection. Due to this requirement careful consideration must be given to the calculation procedure and desired accuracy when commencing the fieldwork. Assuming the depth of the proposed excavation (formation level) is known then each set of spot levels on the grid will form a truncated prism of earth (Fig. 11.20).

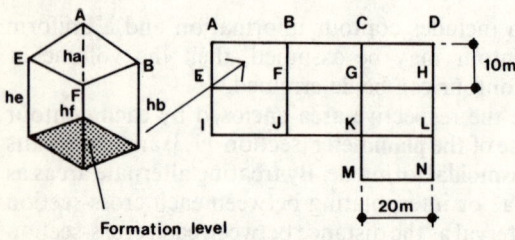

Fig. 11.20 Plan of excavation

The volume of each truncated prism may be determined as follows:

$$\text{Mean depth of prism ABCD} = \frac{h_a + h_b + h_e + h_f}{4}$$

$$\text{Area of square ABCD} = \text{Plan area of grid unit}$$

$$\text{Volume of truncated prism ABCD} = \text{Plan area of grid unit} \times \left(\frac{h_a + h_b + h_e + h_f}{4}\right)$$

This calculation is undertaken for each unit of the grid and the sum will give the total volume required. To aid this calculation Table 11.2 may be used. The example has assumed a formation level of 100 m for the proposed excavation and the spot heights indicated in Table 11.2 above a selected datum.

$$\text{Total volume} = \frac{106.71}{4} \times (10 \times 20)$$

$$= \underline{5335.5 \, \text{m}^3}$$

The accuracy of the procedure may be improved by forming the grid of triangles instead of squares or rectangles, when the volume of each

Volumes from spot height

Table 11.2 Grid levels

1. Station	2. Height above datum	3. Formation level	4. Depth of excavation (3 − 2)	5. No. of rectangles	4 × 5
A	105.62	100.00	5.62	1	5.62
B	104.41	100.00	4.41	2	8.82
C	102.32	100.00	2.32	2	4.64
D	101.05	100.00	1.05	1	1.05
E	107.21	100.00	7.21	2	14.42
F	105.91	100.00	5.91	4	23.64
G	103.60	100.00	3.60	4	14.40
H	102.19	100.00	2.19	2	4.38
I	104.14	100.00	4.14	1	4.14
J	103.62	100.00	3.62	2	7.24
K	102.91	100.00	2.91	3	8.73
L	103.72	100.00	3.72	2	7.44
M	101.21	100.00	1.21	1	1.21
N	100.98	100.00	0.98	1	0.98
				Σ	106.71

trapezoid will be:

$$\text{Volume of trapezoid} = \text{Plan area of triangle} \times \left(\frac{h_a + h_b + h_f}{3}\right)$$

The calculations may be undertaken by adopting the above procedures, although the number of calculations will be increased.

In many applications of excavation work, it is desirable to identify the depth of excavation at which the **cut** or excavated material will balance the **fill** required to form a flat formation level, without any transportation. The required formation level may be determined where a grid has been plotted covering the survey area, with spot heights recorded at each grid intersection, from the following expression:

$$\frac{\text{Mean}}{\text{depth}} = \left(\frac{\Sigma \text{spot height} \times \text{grid weighting}}{4}\right) \div \frac{\text{Number of}}{\text{grid squares}}$$

Note: Where the grid is formed from triangles then the **denominator** in the above expression will be 3 not 4.

To identify the area of proposed excavation, to satisfy the above criteria, the location of the above contour must be plotted on the survey. This is achieved by interpolation between spot heights, using the procedures described in section 8.7.1. The area above this contour will then be excavated to the mean depth and the material removed to the area below the mean depth.

Area calculation

11.12 EXERCISES

1. Calculate from the following data, the area of land between a chain line and a field boundary:

Chainage	0	10	20	30	40	50	60
Field boundary	2.30	3.25	3.85	4.10	4.05	3.65	3.85
Chainage	70	80	90	94.5			
Field boundary	3.20	2.15	0.85	0.00			

(RICS)

2. Calculate the volume of excavation required for the following drain run, assuming vertical sides and a trench width of 1 m.

Chainage	0	15	30	45	60	70	90
Formation depth	14.25	15.01	15.18	15.34	15.50	15.66	15.82
Ground level	18.25	18.00	17.85	18.25	18.75	18.75	18.50

3. The table below gives the metric chainage and offsets from a chain line to a fence. Calculate the area of land between them using:
   (a) Simpson's rule
   (b) Trapezoidal rule
   (c) The average ordinates rule

Chainage	0	10	20	30	40	50	60	70	80
Offset	5.2	7.4	9.0	9.6	9.2	8.4	6.8	3.8	0.4

(RICS)

12 SETTING OUT

12.1 INTRODUCTION

Setting out is the procedure of locating pegs, levels or other means of identification in order that a true interpretation with regard to position, size, shape, level and verticality of a proposed physical feature may be made in relation to its surroundings. These actions are undertaken prior to excavation, tunnelling, building or civil engineering works and are a continuous operation throughout the construction activity. Setting out will demand constant updating, checking and revision from the initial location of the feature to the detailed control demanded when locating specific elements. The best results can only be achieved if recommended procedures are followed which reflect the needs of everyone who will work from the controls provided. To satisfy these demands a range of information may be available, depending on the type of construction activity and stage of work. This will include, any or all of the following:

1. The site plan and setting out drawings.
2. Detailed drawings of the elements of the project.
3. Initial field survey data.
4. Work schedules and location drawings.
5. Architect's instructions.

Before undertaking any setting out activity it is important that the above information is checked and cross-referenced to identify any errors that require further clarification. Once the accuracy of the information has been verified setting out may commence and the procedures adopted in the field will depend on:

(a) the accuracy required at each stage of the construction activity;
(b) the number and characteristics of the feature to be constructed;
(c) the information available.

Setting out

12.2 SETTING OUT EARTHWORKS

12.2.1 Drains and sewers

The location of drains and sewers is fixed from the position of the manholes and appliances being served, therefore it is important that these are accurately located before fixing the controls of the trench excavation. Where the available information fails to identify the exact position of the manholes, these must be agreed from a visual inspection of the ground features, noting the location of possible obstructions and gradients. The location of the manholes on the ground are identified by colour-coded pegs driven firmly into the ground and the location checked by measuring the distance between each peg is then **referenced** and the position **fixed** by triangulation from nearby physical objects so that they may be replaced if disturbed during the construction work.

Once the extent of the proposed excavation has been identified and checked, **line pegs** are placed at 20 m intervals along the length of the drain run, to guide the excavation work and their location and reference is recorded. With mechanical excavation **offset pegs** (Fig. 12.1) are set out at a convenient distance from the centre line on the opposite side to the spoil, since these activities will dislodge any line pegs.

The depth of the excavation will be controlled by **sight rails** placed at each end of the excavation so that a sight line from the **top** of the horizontal rail is parallel to the **gradient** of the sewer. Where mechanical excavation is proposed, the sight rails must be well clear of the trench run and clearly referenced. The depth of the excavation relative to the sight line is controlled by an upright timber member with a cross-head on top which is called a **traveller**. This is used to control the level to the inside of the pipe or channel, referred to as the **invert level** and consequently the depth of the excavation as illustrated in Fig. 12.1.

The top edge of the sight rail is located after calculating the desired length of the traveller (multiples of 250 mm), taking account of the proposed invert level and the need for a comfortable viewing height over the sight rail.

To improve the accuracy of sight rails during the excavation work further sight rails may be provided along the trench run to enable a quick check of any disturbances to be identified.

Example 12.1

A drain is to be laid from a manhole at Station A, 27.65 m (a.o.d) having an invert level of 26.35 m (a.o.d) to a new manhole 75 m distant. The drain is to be laid at a gradient of 1:40 determine:

(a) invert level of the new manhole;
(b) if the sight rail at manhole A is 1.25 m above ground level, calculate the required length of the traveller.

Setting out earthworks

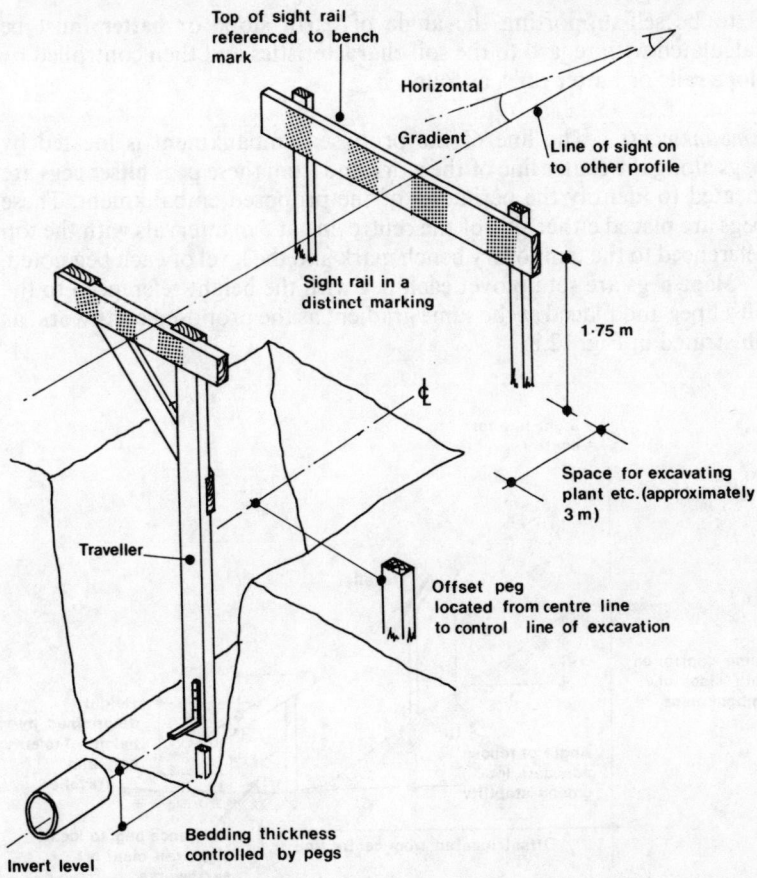

Fig. 12.1 Use of sight rails

(a) invert level of manhole A:

Rise = $\frac{1}{40} \times 75 = 1.885$ m

Invert level = $26.35 + 1.88 = 28.23$ m

(b) Length of traveller = $(27.65 - 26.85) + 1.25$

$= 0.8 + 1.25$

$= \underline{2.05 \text{ m}}$

12.2.2 The control of embankments and cuttings

The control of large earth excavations to provide a level plane for future construction operations are undertaken with sight rails and travellers, similar to the procedures outlined in section 12.2. Where the earthwork

Setting out

is to be self-supporting the angle of earth slope or **batter** must be calculated with regard to the soil characteristics and then controlled by **slope rails** or batter rails, as follows:

Embankments. The line of the proposed embankment is located by pegs along the centre line of the work and from these pegs **offset pegs** are located to identify the perimeter of the proposed embankment. These pegs are placed either side of the centre line at 5 m intervals with the top referenced to the temporary bench mark and the level of each peg noted.

Slope pegs are set up over each peg with the height referenced to the offset peg and placed at the same gradient as the proposed earthwork, as illustrated in Fig. 12.2.

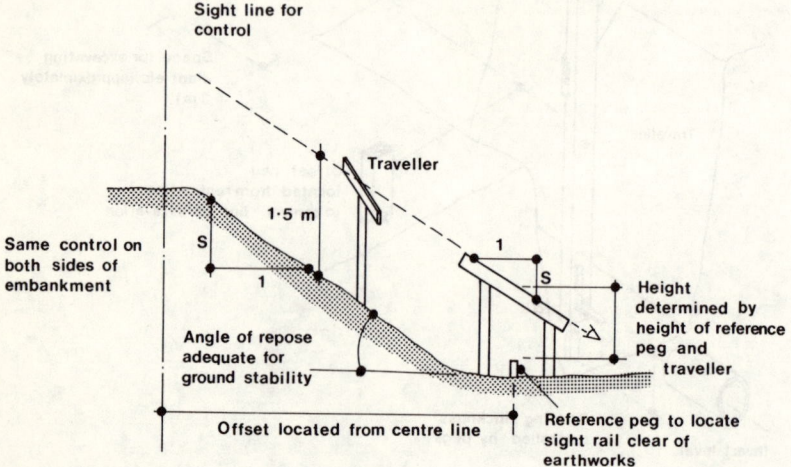

Fig. 12.2 Control of embankments

This procedure is also followed for the control of deep excavations where the side slopes are to be self-supporting. When the height of the earthwork exceeds 5 m further sets of profiles may be constructed on the slope.

Cuttings. The control of cuttings requires a series of sections of the proposed work to be plotted, particularly on a sloping site to determine the perimeter of the excavation at ground level. From the information provided by these sections it will be possible to locate the offset pegs at ground level. These are located behind the limit of the proposed work to control the excavation. After referencing each offset peg to a temporary bench mark on the site, **slope rails** are erected to control the line of the proposed **batter**, as illustrated in Fig. 12.3.

Setting out buildings

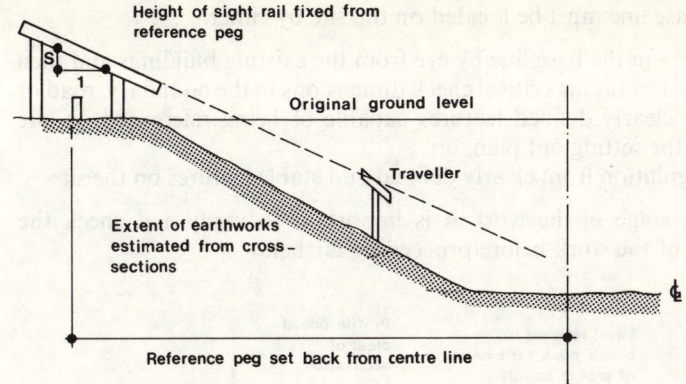

Fig. 12.3 Control of cuttings

12.3 SETTING OUT BUILDINGS

12.3.1 Introduction
The location of a building or a group of buildings on a site will be influenced by:

1. The accuracy required.
2. The layout and characteristics of the building or buildings.
3. The control and setting out data available.

The acceptable cost of the control procedure and the information provided for the location of a building must be related to the required construction tolerances. For a small site it is usually considered most economical to set out from a **base line** which has been clearly identified on the setting out plan and the position fixed on site by pegs.

12.3.2 Setting out from a base line
This procedure is suitable for a small site, eg a dwelling where the site is clear of any physical obstruction. The **base line** must be capable of accurate location on the site from the setting out information provided and the features of the proposed construction are referenced to this base within the tolerances demanded by the project. When existing buildings bound the site, the front face of an adjacent building that has been located on the planning authorities' **building line** may be used for the base line location. When the location of the building line cannot be located on site, an arbitrary datum, that is capable of being identified on the setting out drawings may be used. In each case the base line must be accurately located by pegs, placed in the ground clear of site traffic and the construction work as illustrated in Fig. 12.4.

Setting out

The base line must be located on the site by either:

(a) ranging in the base line by eye from the existing buildings and then verified by taking critical check dimensions to the boundary, road or other clearly defined features capable of being referenced on site from the setting out plan; or:
(b) triangulation from clearly defined and stable features on the site.

At this stage of the work it is important to verify and check the accuracy of the work before proceeding further.

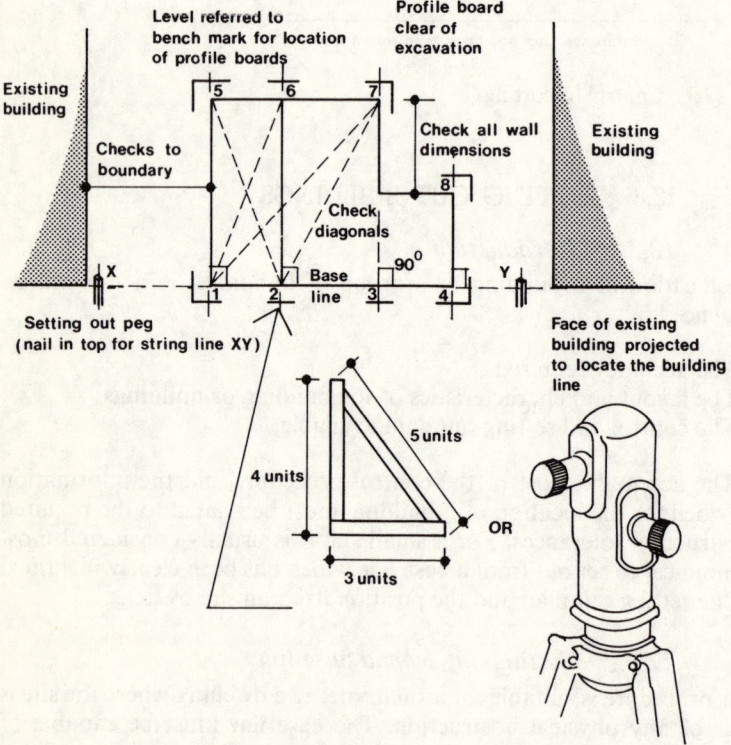

Fig. 12.4 Setting out from a base line

After establishment of the base line illustrated in Fig. 12.4 pegs with nails driven in the centre are located at positions X and Y, clear of the construction work. Using a steel tape stations 1, 3 and 4 are located to fix the **outer face** of the external walls and finally peg 2 to locate the **centre line** of the internal wall. These pegs are located by taking the cumulative dimensions between each peg, all referenced to station X or Y. These are then verified by noting the intermediate dimensions between each peg.

The side walls and internal walls are located by pegs 5, 6, 7 and 8 placed at right angles to the base line, by one of the following procedures:

(a) the optical square (Ch. 3);
(b) a site square which is an optical instrument consisting of two telescopes fixed at 90° to each other and mounted on a tripod, located over the station (Fig. 12.4);
(c) 3:4:5 triangle constructed with a measuring tape as discussed in Ch. 3;
(d) a builder's square constructed on site from timber with sides in the relationship 3:4:5.

From these offsets the rear walls are located by measuring distances 1–5, 2–6, 3–7, and 4–8 with a steel tape and the location of each peg verified by measurement of 5–6, 6–7, 5–7 etc. This work must now be carefully checked before proceeding further by measuring all the **diagonals** from the base line and checking the results with the setting out drawings.

Once the location of the building has been fixed by wooden pegs and verified with regard to the location on site, profile boards may be erected to control the foundation excavation.

12.3.3 Setting out the profile boards

These are erected to provide control of both the **position** and **level** of the subsequent foundation construction. The corner pegs to locate the external face of the wall will be destroyed on commencement of the excavation work, therefore, profile boards are set up well clear of the proposed excavation on an extension of the setting out lines (Fig. 12.5).

In order that **level** control of both the excavation and the concrete work can be undertaken, the horizontal profile board must be erected at a fixed height relative to the proposed structure, eg structural floor level (SFL). This is achieved by referencing the top of the rail to the temporary bench mark placed on site to control all the levels. When transferring the level to the profile board it is important to set the instrument **mid-way** between the temporary bench mark and the proposed location of the profile board to eliminate instrument errors requiring temporary adjustment, this is illustrated in Fig. 12.5.

Using string lines located over the two corner pegs of the wall the position of the external wall face may be transferred to the profile board and identified by a saw cut on the upper edge of the horizontal board. This is undertaken for each wall, and where the corner pegs are too low for direct alignment of the string line a plumb bob may be used. The nails or saw cuts on the profile boards used to identify the location of the external wall face provide a datum from which the foundation width and wall thickness may be set out. The location of the proposed excavation on plan is controlled by running lime, cement or sand lines between the profiles to identify the perimeter of the proposed excavation.

Setting out

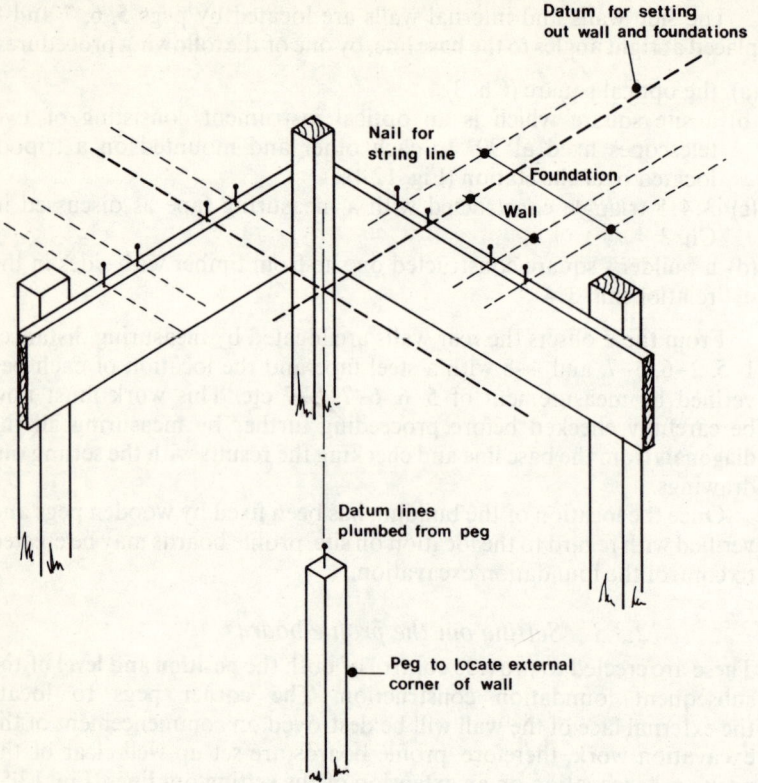

Fig. 12.5 Profile boards

The depth of the excavation is controlled by using the profile board as a sight rail in conjunction with a traveller using the procedures described in section 12.1.

After completion of the foundation concrete and the level of the upper surface of the footing has been verified, the position of the proposed wall must be accurately located. This is achieved by placing a string line between each profile board to identify the inner and outer wall face in order that the location may be transferred to the foundation by means of a plumb line.

12.3.4 Setting out from a grid

Selection of the grid. A development which includes a number of buildings or a large building which is uniform on plan, eg a steel-framed structure which demands fine tolerances when locating the components and detailed co-ordination between members of the construction team, will benefit from the use of a setting out grid. The setting out drawings

204

Setting out buildings

may include a grid that can be clearly related to the site features; alternatively, a grid may be superimposed on the drawings, checking first that it is square and dimensionally accurate. The selection of this grid will be influenced by:

(a) the provision of clear lines of sight when setting out the grid and the ability to reference the proposed features to the grid;
(b) the orientation of the proposed structure;
(c) the possibility of adopting the standard modules used in the design of the structure;
(d) the need to establish main grid check points clear of site traffic and any possible disturbances;
(e) the need for any secondary grids required for control purposes during the construction period;
(f) economy of measurement consistent with the need for checks on the accuracy of the work;
(g) limitations imposed by the site boundary and other physical features.

Having established the setting out grid on the site plan and associated setting out drawings, the reference frame must be located on the site (Fig. 12.6) and the intersections of the grid (**key points**) established. A permanent reference to the **key points** must be made by means of **reference pegs** located in a position free from possible disturbance. These pegs must be protected by means of a concrete base and possibly a fence with a triangulated fix made to other permanent locations. Where a location free from possible disturbance cannot be located the reference may be provided by a steel plate sunk below ground and protected by sand with a cover placed over. These permanent references may also be used as a temporary bench mark where **both** accurate level and plan location are important, including pile caps and steel stanchions.

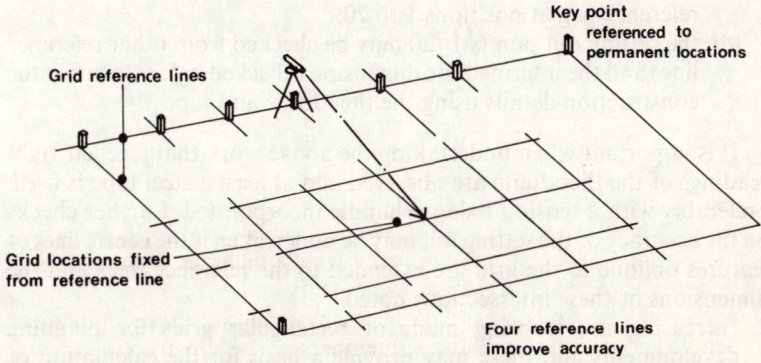

Fig. 12.6 Grid reference frame

Setting out

Setting out the grid. Once the grid reference lines and key points have been established the location of the column, pile centres and other features may be fixed by:

1. Grid references located using a theodolite and steel tape as illustrated in Fig. 12.7

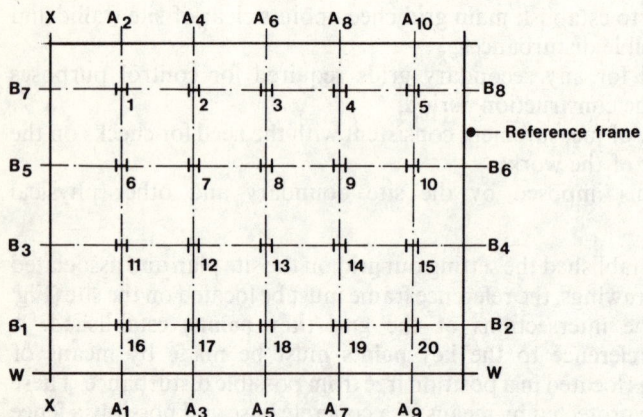

Fig. 12.7 Grid location

 (a) set out Stations $A_{1,3,5,7,9}$ along the reference line W–W;
 (b) set out stations B_1 to B_7 along reference line X–X;
 (c) the centre line of each station can then be located from the above reference lines using the theodolite and steel tape, placing a reference peg at positions 1 to 20;
 (d) the setting out points 1–20 may be checked from other reference lines and the intermediate dimensions checked before locating the construction details using the theodolite and tape.

It is important when undertaking the above work that face left/right readings of the theodolite are observed and at least a steel tape is used, preferably with a tension balance handle incorporated. Further checks on the accuracy of the setting out may be undertaken if the **centre lines** of features oblique to the grid are extended to the reference lines and the dimensions of these intersections noted.

2. Increasing use is being made of **rectangular grids** for planning developments and these may provide a basis for the calculation of **rectangular co-ordinates** to fix stations with regard to the frame.

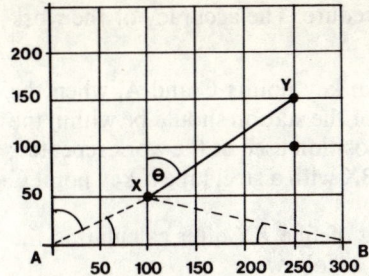

Fig. 12.8 Grid co-ordinates

(a) To locate station X from **key point** A in Fig. 12.8.

	Partial co-ordinates	
Station X	50	100
Station A	0	0
Difference	50	100

$\text{BAX} = 50/100 \tan^{-1} = 26°33'54''$
$\text{AX} = 100/\cos 26°33'54''$
$= 88.44 \text{ m}$

(b) To locate station X from **key point** B.

	Partial co-ordinates	
Station X	50	100
Station B	0	300
Difference	50	200

$\text{ABX} = 50/200 \tan^{-1}$
$= 14°02'10''$
$\text{XB} = 200 \cos 14°02'10'' = 194.03 \text{ m}$

By locating a theodolite over Stations A and B each focused onto the other station and then rotated through the respective angle BAX at Station A and ABX at Station B, position X may be located at the intersection of the telescope sightings (see Fig. 12.8). Station Y may then

Setting out

be located by adopting the same procedure. The accuracy of the work may be verified by:

(a) adopting the same procedure from key points C and A, when the discrepancy between the location of the station should be within the tolerance required and the mean position used **or** the work repeated;
(b) Measuring the distances AX and BX with a steel tape if key point C cannot be located;
(c) Measuring the length and bearing of line XY after calculating the true values from the co-ordinates, as follows:

	Partial co-ordinates	
Station Y	250	150
Station X	50	100
Difference	100	50

$$XY = \sqrt{100^2 - 50^2}$$
$$= 111.80 \text{ m}$$

$$\text{Bearing} = 50/100 \tan^{-1}$$
$$= 56°18'56''$$

The procedure of calculating the length and bearing of the line joining two stations whose co-ordinates are known is termed **calculating a join** and has a wide range of applications, including the following:

(a) the setting out of stations at known co-ordinate positions;
(b) checking the co-ordinates of pegs to determine whether they have been disturbed;
(c) locating the reference framework of a setting out grid from two known stations.

12.4 THE VERTICAL CONTROL OF A STRUCTURE

The procedures for controlling the vertical alignment of a structure, both above and below ground level, will depend on the tolerances required and the operating conditions. The application of plumb lines for the transfer of a line to the bottom of an excavation has been discussed in section 1.3.3 and the accuracy may be improved by the use of wire whose oscillation is damped by immersion of the weight in a bucket of oil.

The demand for accurate plumbing of the elements of a structure and the transfer of control grids from the ground to higher levels within a building for dimensional and positional control has led to the development of special instruments for vertical alignment. The high cost of these

The vertical control of a structure

instruments compared with the simple procedures outlined overleaf is justified by the improved accuracy and speed of operation. The instruments available include:

The optical plummet. This is based on the optical plummet used in conjunction with the theodolite (section 5.8.2) and is often manufactured to be mounted on the same tribrach for accurate centring. It consists of a small telescope which incorporates a prism to deflect the line of sight in a vertical direction. There are a range of instruments available but most include a self-levelling device to increase the accuracy from 1 in 10 000 to 1 in 40 000 or above with a range up to 100 m. Attachments are also available for locating on the telescope of the theodolite and when properly located with the sight line of the optical plummet focused over the station will define a vertical line of sight upwards and in line with the plummet.

The auto plumb. This compact device is manufactured to order by Messrs Hall & Watts and incorporates two eyepieces, the lower acting as the conventional optical plummet sighting downwards, while the upper eyepiece sights upwards and incorporates an automatic level compensator. This compensator defines a vertical line when it is set up level but a micrometer drum will tilt the prism, enabling the sight line to be offset from the vertical. This facility enables the deviation of a structure from the vertical to be measured and any lateral displacement to be calculated (Fig. 12.9). The instrument is located in a vee slot on the

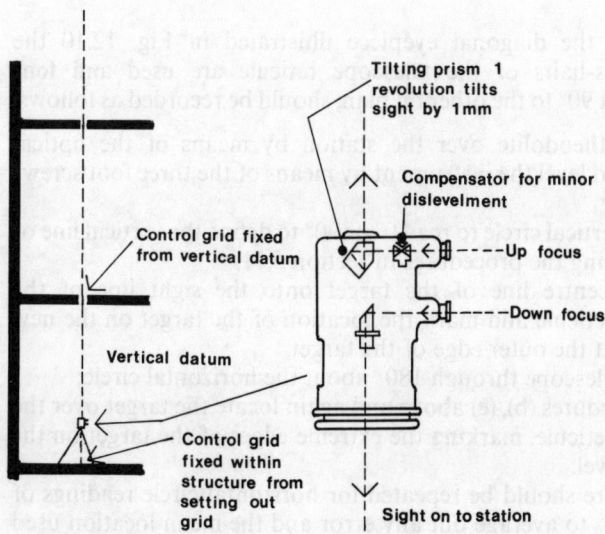

Fig. 12.9 Watts autoplumb (courtesy Hilger and Watts)

Setting out

mounting and there are four positions on plan for locking the instrument. By taking at least two readings each at 180° to the other on plan it is possible to achieve a centring accuracy of = 0.3 mm in 40 m.

Vertical control using the theodolite. Even though the conventional theodolite has a telescope which is capable of being rotated about the **vertical** plane, it is restricted by the available space adjacent to the reference station and the maximum elevation that it may be transited (approximately 25°) with accuracy while allowing the vertical circle to be observed with ease. These restrictions are reduced if a diagonal eyepiece is fitted in place of the conventional eyepiece since this improves the observation procedures when the telescope is elevated above 25°.

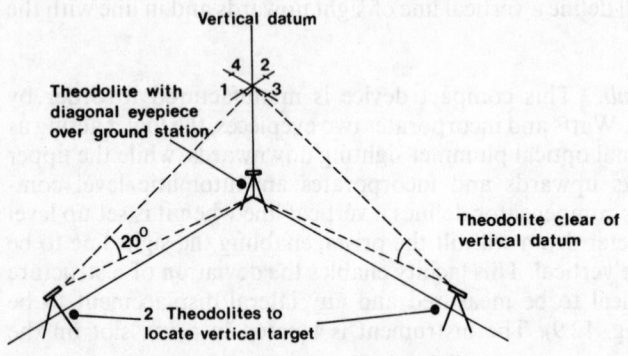

Fig. 12.10 Vertical control using the theodolite

When using the diagonal eyepiece illustrated in Fig. 12.10 the horizontal cross-hairs of the telescope reticule are used and **four** readings, each at 90° to the other on plan, should be recorded as follows:

(a) set up the theodolite over the station by means of the optical plummet and level the instrument by means of the three foot screws (section 5.8);
(b) centre the vertical circle to read 0° or 90° to define the vertical line of sight, following the procedures in section 5.11;
(c) locate the centre line of the target onto the sight line of the horizontal reticule and mark the location of the target on the new floor level, at the outer edge of the target;
(d) rotate the telescope through 180° about the horizontal circle;
(e) repeat procedures (b), (c) above and again locate the target over the horizontal reticule, marking the extreme edges of the target on the new floor level;
(f) the procedure should be repeated for horizontal circle readings of 90° and 270° to average out any error and the mean location used for future setting out.

12.5 EXERCISES

1. A drain is to be laid from a new manhole to an existing manhole 60 m distant. The invert level of the existing manhole is 46.05 m and the drain is to a fall from the new manhole to the existing manhole of 1:40. The ground level varies from 49.05 m near the new manhole to 47.60 m near the existing manhole. Calculate the invert level of the new manhole and describe how you would mark out the line of the drain with sight rails ready for excavation. (RICS)
2. Describe how you would set out a detached dwelling on an open site; mention particular procedures that would ensure the necessary accuracy and permanency in your setting out. (RICS)
3. Discuss the setting out of a steel-framed building and, using sketches as required, describe how squareness of plan, levelness of foundations and verticality of uprights are achieved.

13 THE MEASUREMENT OF EXISTING BUILDINGS

13.1 EQUIPMENT

The surveyor is often required to undertake a measured survey of an existing building for alteration, maintenance or dilapidations reports. The procedures adopted reflect the principles outlined in other chapters for the measurement of land and require the following basic equipment:

(a) measuring tape for the general measurement of dimensions, steel tapes are used for greater accuracy;
(b) folding measuring rod, since this has greater rigidity than the flexible steel rule when measuring elevations, etc;
(c) chalk;
(d) sketch book, which should contain some squared paper to assist the accuracy of the sketches;
(e) surveyor's chain when external detail must be located by offsets, etc;
(f) plumb line for obtaining vertical dimensions and use as a datum for measuring elevations, sections, etc.

13.2 GENERAL PROCEDURE

The objectives of the survey will direct the procedure to be followed, the data to be recorded and the scale of the final plotting although the following activities will be undertaken:

Brief stage.   Identify objectives of the survey and determine what information is required, taking special note of any unusual requirements.

Reconnaissance stage.   Before proceeding to measure the required features a thorough reconnaissance of the building and locality should be carried out. Consideration should be given to the order of the survey and any special problem identified.

Sketch stage. Sketch the following, in order, after estimating the size and extent of the detail required:

(a) plan of the site;
(b) plan of each floor;
(c) roof plan;
(d) all elevations;
(e) all relevant sections and details.

Each sketch should be on a separate sheet, each identified with the address of the building, data, title of sketch and surveyor's name.

Measurement stage. Take all measurements, noting them clearly on the appropriate sketch.

Collection of other information. Collect all other information, eg positions of services, description of construction details and other characteristics relative to the survey objectives.

Check stage. Before leaving the site check over the notes, sketches, details, etc, to ensure that nothing has been omitted and walk over the survey area, checking that nothing has been omitted.

Plotting stage. Plotting should be undertaken as soon as possible to ensure that all the information has been collected.

13.3 BOOKING PROCEDURES

13.3.1 Dimensions

By holding one end of the tape at a fixed location and running the tape along the direction to be measured all the dimensions can be related to the chosen location. The method follows the same procedure as reading a chain, section 3.5, and are called **running measurements** (Fig. 13.1).

This procedure is recommended where a series of measurements are to be recorded since an error in booking one dimension will not affect the overall dimension.

The person booking the readings will usually hold the end of the tape and repeat each reading booked as a check.

13.3.2 Plans

These should identify the shape and location of each property within the boundary of the survey. The survey procedures should follow the principles of the chain survey outlined in Ch. 3 with all the main features located by triangulation (Fig. 13.2) and capable of being plotted

The measurement of existing buildings

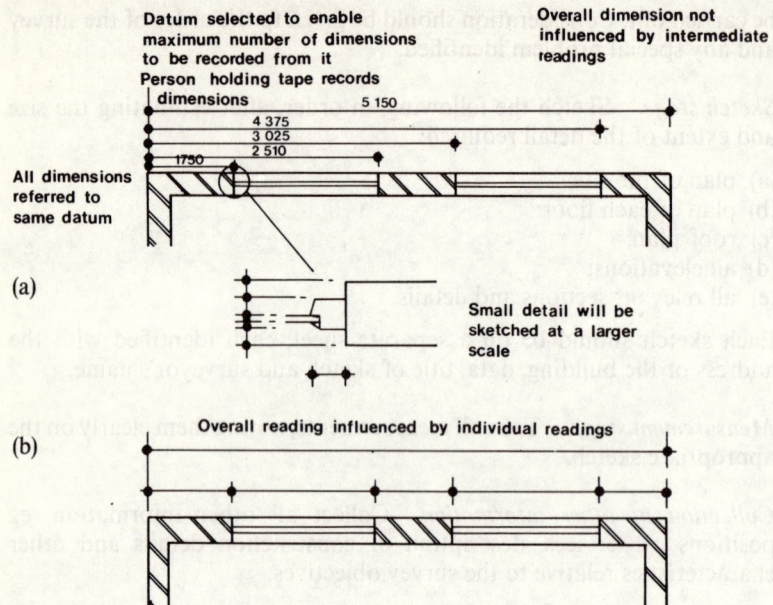

Fig. 13.1 Measurement procedure (a) running measurements (b) individual dimensions

especially the corners of the building. Care must be taken to:

(a) measure ties for triangulation to help with the plotting;
(b) relate floor levels inside the building with outside ground level;
(c) measure heights from sills, etc in relation to outside ground level.
(d) locate all gullies, manholes, drain runs, rainwater pipes and soil pipes.
(e) measure the thickness of walls at all door and window openings.

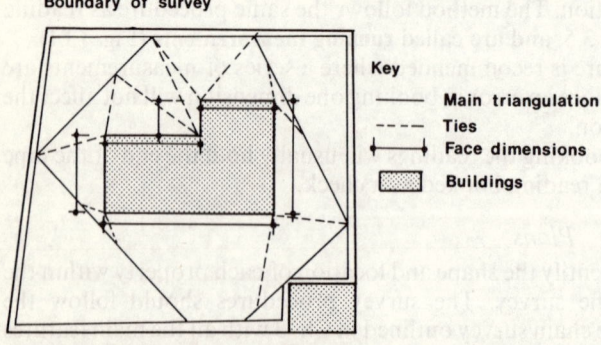

Fig. 13.2 The block plan (RICS)

Each room should be taken in turn once an outline sketch has been made, working in a clockwise direction. Draw over the initial guide lines and identify all the relevant features, doors, windows, recesses, columns, bays, etc. The direction of the joists, floor boards, etc should also be indicated with a note of the materials used (Fig. 13.4).

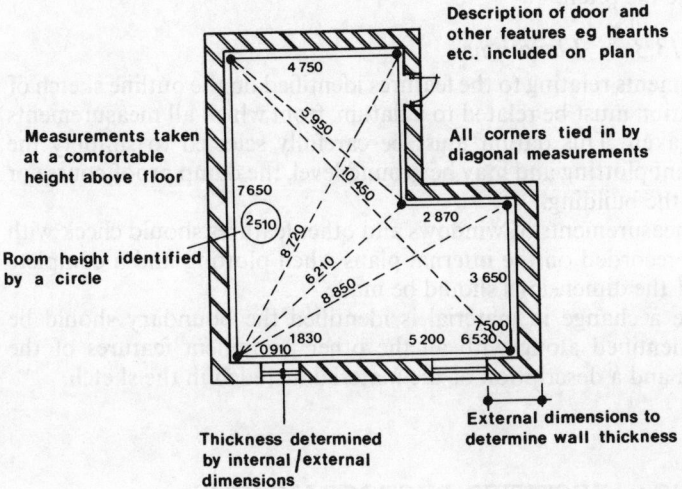

Fig. 13.3 Internal plan detail

After running measurements have been recorded diagonal measurements may be taken to ensure accurate plotting. Where the detail is too small to be clearly identified a larger sketch can be included providing it is clearly referenced. Many projections may prove difficult to accurately measure and the procedures shown in Fig. 13.4 may be adopted.

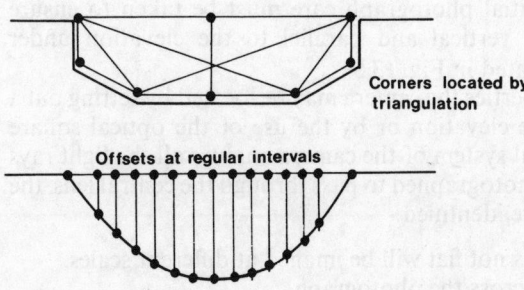

Fig. 13.4 The recording of bays

The measurement of existing buildings

The location of openings or internal walls should never be assumed once they have been identified on a plan and complete measurements should be taken for all subsequent plans. This assumption may have to be made with the roof plan when access is not possible and an appraisal may have to be made from external observations along with measurements estimated by counting the courses of brickwork or tiles to determine the pitch.

13.3.3 Elevations

Measurements relating to the features identified on the outline sketch of the elevation must be related to a datum, from which all measurements will be taken. This datum must be carefully selected to simplify the subsequent plotting and may be ground level, the damp proof course or eaves of the building.

The measurements of windows and other features should check with the data recorded on the internal plans when plotting and a complete record of the dimensions should be made.

Where a change in material is identified the boundary should be clearly identified along with all the other prominent features of the elevation and a description of the material included in the sketch.

13.4 RECTIFIED PHOTOGRAMMETRY

Photogrammetry is a technique of obtaining a detailed record of the elevation of a building from a photograph and has the following advantages:-

1. The equipment is relatively cheap.
2. The photographic image may provide far more information than a line drawing.
3. The photograph provides an accurate record of the elevation.
4. The procedure can be rapidly undertaken especially when linked to a computer and graph plotter.

When taking the initial photograph care must be taken to ensure that the film plane is vertical and parallel to the elevation under consideration as illustrated in Fig. 13.5.

To ensure these properties the camera may be located by setting out a 3:4:5 triangle from the elevation or by the use of the optical square (Ch. 3). Since the optical system of the camera enables all the light rays from the object being photographed to pass through the central lens, the following limitations are identified:

1. An elevation which is not flat will be imaged at different scales.
2. The scale will vary across the photograph.
3. The enlarger system may introduce inaccuracies.

4. The camera should not normally be tilted to include the upper sections of tall buildings, although specialists in this field can remove tilt errors when developing.

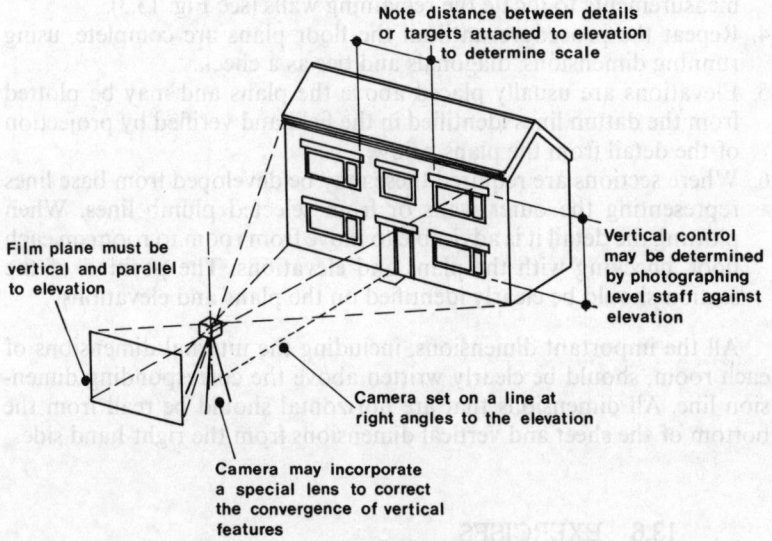

Fig. 13.5 Principles of rectified photography

Control measurements are taken to identify a scale from the photographs and at least one vertical and one horizontal distance should be recorded.

The resulting photographs are enlarged to the correct scale and printed on to transparent film or A4 size sheets. The procedures claim to provide a relatively high order of accuracy with a maximum error of ± 40 mm at a scale of 1:50.

13.5 PLOTTING

The plans, sections and elevations are usually plotted in the same order as they were recorded using orthographic projection to provide an accurate and detailed record of the building. The procedure for plotting may follow the stages below:

1. Draw the ground floor plans first. These are usually located in the bottom left-hand area of the sheet with the front of the building facing the bottom of the sheet.
2. Set out a line to represent the largest external wall to form a base line on which the remaining features can be triangulated. Indicate the

The measurement of existing buildings

thickness of this wall and identify the position of window openings, reveals, etc, checking the plan dimensions with those obtained for the elevations.
3. Plot all the walls and features relative to this wall using the diagonal measurements to locate the remaining walls (see Fig. 13.3).
4. Repeat the procedure until all the floor plans are complete, using running dimensions, diagonals and ties as a check.
5. Elevations are usually placed above the plans and may be plotted from the datum lines identified in the field and verified by projection of the detail from the plans below.
6. Where sections are required these may be developed from base lines representing the outer walls or from selected plumb lines. When plotting the detail it is advisable to move from room to room on each floor, checking with the plans and elevations. The position of the section should be clearly identified on the plans and elevations.

All the important dimensions, including the internal dimensions of each room, should be clearly written above the corresponding dimension line. All dimensions that are horizontal should be read from the bottom of the sheet and vertical dimensions from the right-hand side.

13.6 EXERCISES

1. Alterations have to be carried out on an existing two-storey house. There are no records or plans of the building. A survey is to be made to provide the necessary information for plans to be drawn up for the alteration to be made. State what building plans, elevations, etc should be provided for the contractor. Describe the technique of measuring up the room shown in Fig. 13.6. (RICS)

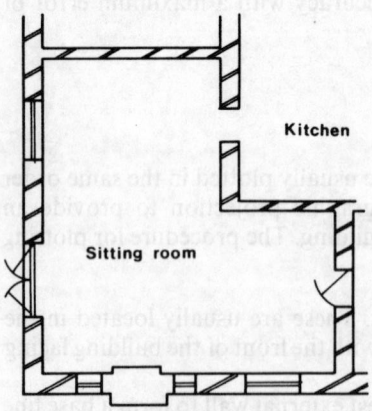

Fig. 13.6 Sitting room (RICS)

Exercises

2. Figure 13.7 shows an outline of an existing single-storey building, approximately 8 metres × 5 metres overall, which it is proposed to convert into a dwelling. As no drawings are available from which to prepare a scheme, it is necessary to measure the existing building in order to produce a plan of it. Show by sketches the information to be recorded on site and how measurements would be taken along A–B. State the principal items of equipment required for the survey.

(RICS)

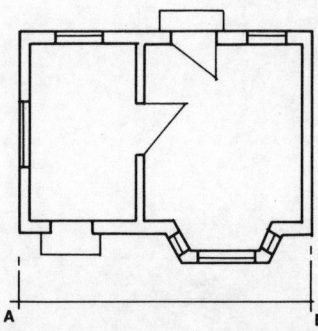

Fig. 13.7 Single storey building (RICS)

3. Give an account of the procedures to be followed in measuring a two-storey pitched roof detached house with a separate garage, in order to produce a site plan, ground and first floor plans, elevations and cross-sections. Illustrate your answer with clear sketches and include information regarding:

    (a) the equipment required for the survey;
    (b) the scales at which the various drawings will be produced; and
    (c) the accuracy with which measurements should be made in order to produce drawings at the required scale. (RICS)

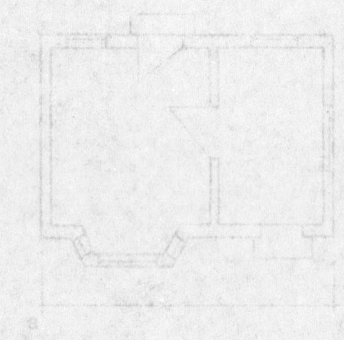

II STATISTICAL ANALYSIS

INTRODUCTION

This section deals with the basic principles of statistical methods.

The goal is to introduce the ideas and concepts which are fundamental to the understanding of modern statistics.

Statistical data and statistical techniques play essential roles in many decision-making and forecasting situations and it is becoming more apparent that anyone requiring to understand the implications of much numerical work needs at least an acquaintance with the subject of statistics.

It is assumed that the reader has a fairly limited mathematical background and some mathematical detail is omitted. It is important that the meaning and implications of basic concepts are understood, but it is hope to avoid the reader getting lost in an excessive amount of detail, which may cloud the important issues.

14 DATA AND ITS PRESENTATION

14.1 INTRODUCTION

One of the fundamental aims of statistics is to make sense of a mass of unorganised data. To fully appreciate the information presented by a mass of figures some order or system must be brought about.

In this chapter we look at some of the ways of presenting data. Basically this involves making data more informative, more interesting, more acceptable and also in many cases more manipulative for mathematical purposes.

When collecting information we want to find out about a particular characteristic of a set of people or objects, eg the values of properties, the weights of materials, etc.

The particular characteristic in which we are interested is called the variable because it varies from one member of a group to another.

14.2 ORGANISING DATA: FREQUENCY TABLES

The following figures show the thickness to the nearest mm of twenty-five concrete blocks which a manufacturer has selected from his stock.

98	99	99	102	100
101	102	99	100	99
100	102	101	100	101
103	100	98	98	100
97	100	102	100	103

Data presented in this manner as a collection of single facts are not very meaningful. To make these figures useful it is necessary to organise them in some systematic manner.

Grouping, of data

A good method of doing this is to draw up a frequency (distribution) table, which is so called because it shows the frequency with which the various figures occur.

The way to tabulate these data is to write down, in order, the different thicknesses and count how many times each figure occurs. This counting can be facilitated by the use of tally marks; a stroke being placed against a figure each time it occurs and every fifth stroke being drawn diagonally across the previous four to make groups of five. The number of times each value occurs is called the frequency (designated by the letter 'f') of that value.

Table 14.1. Shows the tabulation results.

Table 14.1 Frequency distribution of thickness of concrete blocks

mm	Tally	Frequency (f)
97	1	1
98	111	3
99	1111	4
100	1111 111	8
101	111	3
102	1111	4
103	11	2
		25

This simple organisation of the figures often provides a first step towards obtaining statistical information.

14.3 GROUPING OF DATA

Sometimes the data we are considering contain a large range of values and it is necessary to collect the data into groups.

The following are the rating valuation assessments of forty residential properties in a certain street.

158	176	188	208	183
190	216	184	184	191
164	174	154	186	178
186	202	179	185	183
177	198	182	174	193
164	197	170	179	162
167	178	173	182	194
192	199	194	177	169

These figures are widely spread out (the lowest value being 154 and the highest 216) and to draw up a frequency table similar to the one in the previous example would produce an unwieldy and impractical table.

223

Data and its presentation

In such a case as this it is customary to group or classify the data. The figures are grouped into what are called class intervals and a frequency for each group is obtained.

The main problem lies in choosing the class intervals to use. This choice is basically a matter of judgement: it really depends upon what we want to do with the data once it has been classified. The class intervals should not be so large as to lose much of the information obtained from the original data, eg if the rating valuations were merely categorised into those below £190 and those £190 and above, much of the relevance of the valuations would be lost. On the other hand, classification into groups with a range of say just two or three marks would defeat the purpose of grouping.

In this example, as a happy medium, the marks have been classified as shown in Table 14.2.

Table 14.2 Frequency distribution of rating valuations

£	Tally	Frequency (f)
150–159	11	2
160–169	⊔⊔⊔	5
170–179	⊔⊔⊔ ⊔⊔⊔ 1	11
180–189	⊔⊔⊔ ⊔⊔⊔	10
190–199	⊔⊔⊔ 1111	9
200–209	11	2
210–219	1	1
		40

This classification enables us to obtain an immediate picture of the distribution of the valuations, ie a clustering of valuations in the £170s, £180s and £190s and fewer properties with valuations at either extreme. Therefore one objective of neatly displaying the information contained in the raw data has already been achieved.

However, one point to be noted which will be important at a later stage, is that some accuracy must by necessity be lost when the figures are grouped, eg from Table 14.2 we can see that five properties had valuations between £160 and £169 but we cannot tell that they were £162, £164, £164, £167 and £169 respectively.

Choice of class boundaries. Several factors were taken into consideration in this example.

1. Each class had to be mutually exclusive and no gaps to be left between classes. This meant that any valuation could be assigned to just one class. It would have been impossible to have classes £150–160, £160–170, etc. Into which class could we place a valuation of £160?
2. Class intervals were chosen that were easy to read and use, eg intervals of 5, 10, or 100 units are mentally easier to cope with.

Grouping of data

3. Equal class intervals were chosen. Generally speaking a frequency table is 'easier' to read if the classes are of equal length. In this example our data are likely to be of most use when categorised as valuations in the £150s, in the £160s etc.

However, this is certainly not a hard and fast rule and circumstances often dictate that frequency tables contain classes of varying length especially at the beginning or end of tables.

4. The first class began with £150. There would have been little point in having classes £130–139, £140–149, for instance, as this would not have added any extra information to the table. The lowest and highest values in the data (ie £154 and £216) were important in deciding on the first and last classes for the table.

It is worthwhile at this point to distinguish between discrete and continuous variables.

A variable is discrete if it can only have a finite number of values, or usually as many values as there are whole numbers. The number of people living in an area, for example, is a discrete variable because it has to be a whole number.

In contrast, a continuous variable is one which can assume any value along a continuous scale. Length, weight and time are all continuous variables, even though in practice the way in which these variables are measured and rounded off (eg to the nearest mm, kg or second) makes them discrete.

If the class intervals in our example were plotted on a scale, there would be gaps between the classes.

```
            159   160                    169   170
        ──────────┤ ├─────────────────────────┤ ├──────
```

This did not really matter in this case because all the valuations were given in whole pounds. If the values had been put to the nearest penny the class limits, ie the smallest and largest values which could fall in a particular class, would have had to have been changed from, for example, £150–159 to £150.00–£159.99 to accommodate all the possible values.

In certain circumstances, we may wish to eliminate the gaps altogether and in that case we may have to define class limits. For the first class the lower class limit would be £149.51 and the upper class limit £159.50; for the second class the lower class limit would be £159.51 and the upper class limit £169.50, etc.

Open-ended classes. When dealing with data in which just one or a few values are much greater (or smaller) than the rest of the values then an open-ended class can be used in the frequency table.

If there had been just one property in our example with a rating valuation of say £283, the simplest way of incorporating this into the

Data and its presentation

table would have been to add one extra class to the existing table by putting

£	f
⋮	⋮
220 and above	1

Similarly, a single relatively low valuation of, say, £108 would have meant an extra initial class such as

£	f
Under 150	1
⋮	⋮

The extra class in both these cases is termed an open-ended class and has only one class limit.

14.4 CUMULATIVE FREQUENCY

It is often useful to arrange the data from a frequency distribution into a cumulative frequency distribution.

In many situations we may wish to know how many values fall above or below certain limits rather than how many fall in a particular class.

Table 14.3 shows a frequency distribution of the marks (out of 100) of 50 students in a surveying examination.

The students may obtain different levels of grading according to whether they obtained certain marks and so we could cumulate the frequencies in two different ways.

Firstly, to construct a cumulative 'less than' distribution, from the first class in Table 14.3 we can see that only 1 student obtained less than

Table 14.3 Frequency distribution of marks in a surveying examination

Mark	f
0–9	1
10–19	2
20–29	4
30–39	4
40–49	7
50–59	15
60–69	9
70–79	3
80–89	2
90–99	3
	50

Cumulative frequency

10 marks: looking at the second class as well, by adding the frequency for the second class to the frequency for the first class, we find that $(2 + 1) = 3$ students obtained less than 20 marks.

The cumulative distribution resulting from the procedure is shown in Table 14.4. Note that the last value in the cumulative frequency column must be the same as the total frequency.

Table 14.4 gives us information on how many students failed to achieve certain marks. It may be of greater interest to cumulate the frequencies beginning at the other end and draw up an 'or more' distribution as in Table 14.5.

There are 3 students with 90 or more marks, as can be seen from the last class in Table 14.5, and if we add the frequency for the penultimate class to this, we find that there are $(3 + 2) = 5$ students who obtained 80 or more marks. This adding procedure going back one class each time gives us the full table.

Table 14.4 Cumulative 'less than' distribution of surveying examination results

Mark	Cumulative frequency
Less than 10	1
Less than 20	3
Less than 30	7
Less than 40	11
Less than 50	18
Less than 60	33
Less than 70	42
Less than 80	45
Less than 90	47
Less than 100	50

Table 14.5 Cumulative 'or more' distribution of examination marks

Mark	Cumulative frequency
0 or more	50
10 or more	49
20 or more	47
30 or more	43
40 or more	39
50 or more	32
60 or more	17
70 or more	8
80 or more	5
90 or more	3

Data and its presentation

We could also have drawn up a cumulative 'or less' table, ie '9 or less', '19 or less', etc, but this would give the same cumulative frequencies as in Table 14.4 and as the pass marks for examination gradings are usually set at the 10 mark points the table shown is preferred.

Similarly Table 14.5 is preferred to a cumulative 'more than' table, ie 'more than 89', etc in this particular example.

14.5 GRAPHICAL PRESENTATIONS OF DATA

Having put data into a frequency table, one common follow-up is to present the data in a graphical form in order that essential features of the frequency distribution can be displayed.

14.5.1 Histograms

One way of representing a frequency distribution graphically is by means of a histogram.

On the graph, the values or measurements are represented on the horizontal scale, and the frequencies on the vertical scale.

Rectangular blocks are then drawn, the bases of which are determined by the values – or class intervals in the case of grouped data – in the frequency table. The heights of the blocks are determined by the corresponding frequencies.

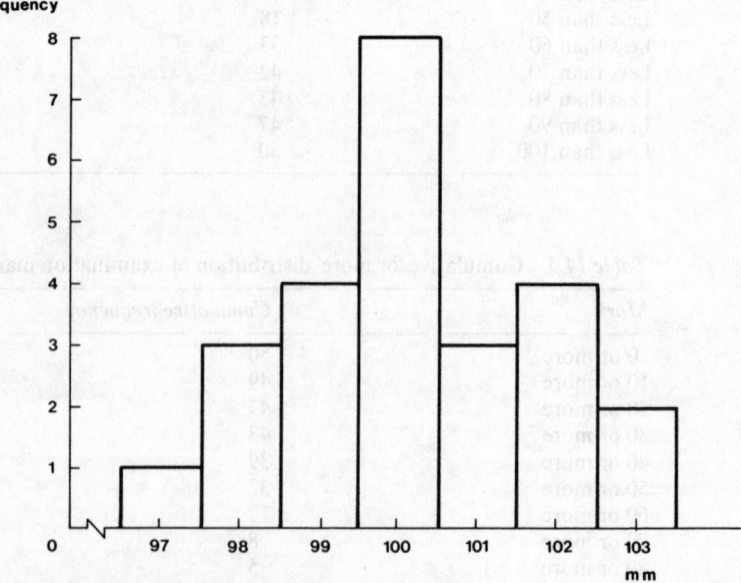

Fig. 14.1 Histogram from Table 14.1

Graphical presentations of data

Figure 14.1 shows a histogram plotted from the information contained in Table 14.1. The centre of each block is set against the value which it represents on the horizontal axis.

In the case of grouped data, on the horizontal scale we can either indicate the class boundaries or, more usually, if we want to make the histogram easier to understand, we can use more approximate key values.

Figure 14.2 shows a histogram plotted from the grouped data in Table 14.2. The width of each block is 10 units on the horizontal scale. To be strictly accurate we should take the class boundaries such as £159.5 and £169.5 as the boundaries of each block, but the presentation is improved here if the values of £160, £170, etc are used instead.

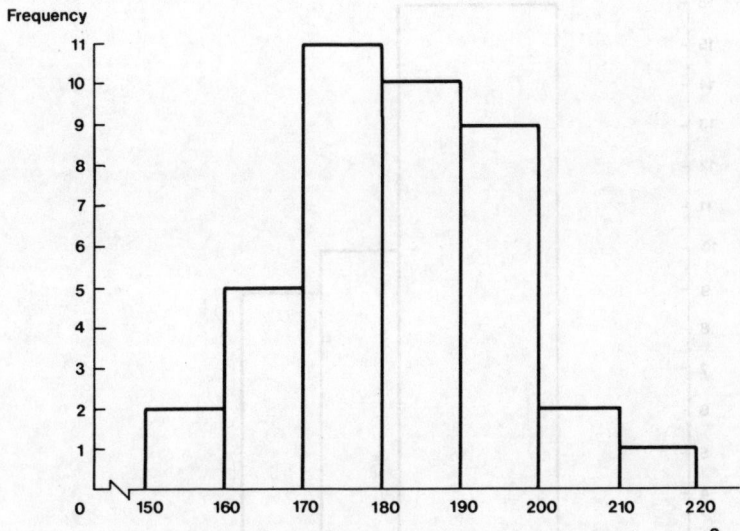

Fig. 14.2 Histogram from Table 14.2

The rule that class intervals should be equal is often broken. Care has to be taken that a histogram is not misleading, as it indeed might be if adjustments are not made in the case of a distribution with unequal class intervals.

Table 14.6 shows the data on rating valuation assessments when all valuations from £160 to £179 are put into one class.

The histogram which corresponds to the frequency distribution is shown in Fig. 14.3. This histogram misrepresents the data in the sense that it probably gives the impression that the majority of the valuations lie in the £160 to £180 range, whereas this is not actually the case.

229

Data and its presentation

Table 14.6 Altered frequency distribution of rating valuations

£	Frequency (f)
150–159	2
160–179	16
180–189	10
190–199	9
200–209	2
210–219	1
	40

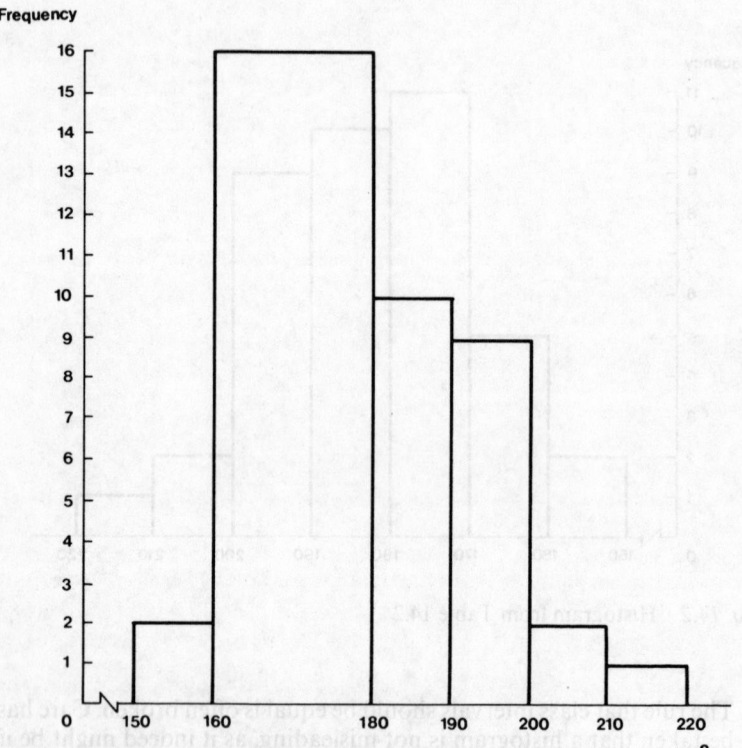

Fig. 14.3 Misleading histogram from Table 14.6

To rectify this misleading picture, we need to adjust the histogram block representing the second class. This involves dividing the frequency of the class by two and then drawing the block to this new height of 8.

Generally, if one class is twice as large as the others we divide its frequency by two; if three times as large, divide by three, etc.

The assumption implicit in this adjustment is that a half of the 16 values in this class fall in the £160–£169 group and the other 8 fall in the £170–£179 group.
The adjusted histogram is shown in Fig. 14.4.

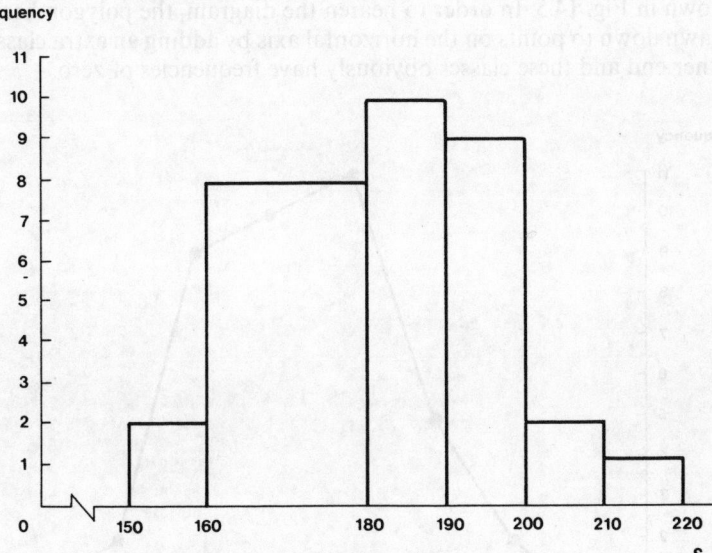

Fig. 14.4 Adjusted histogram from Table 14.6

It is the area of each block which represents the relative importance of each class and not just the height of the block.

Finally it needs to be mentioned that a histogram is not really suitable for dealing with open-ended classes. If it is essential however that an attempt be made to draw such a diagram, the block for the open-ended class is best displayed as being of the same width as the adjacent block and the class limit of the end block labelled 'or more' or 'less than', etc according to the particular requirement.

14.5.2 *Frequency polygons*

Another common form of graphic representation of a frequency table is the frequency polygon.

It can be drawn quite simply by plotting the frequency associated with each value – or the mid-point of the class interval in the case of grouped data – and then connecting these points by straight lines.

If a histogram has already been drawn, the mid-points of the tops of the blocks can be joined together with straight lines to form the polygon.

Data and its presentation

However, it is not necessary to construct a histogram before drawing the polygon, and given a choice between the two types of diagram to represent certain distributions, it is common practice to use a histogram for discrete distributions and a frequency polygon for data where continuity is likely.

A frequency polygon representing the data on rating valuations is shown in Fig. 14.5. In order to neaten the diagram, the polygon here is drawn down to points on the horizontal axis by adding an extra class to either end and these classes obviously have frequencies of zero.

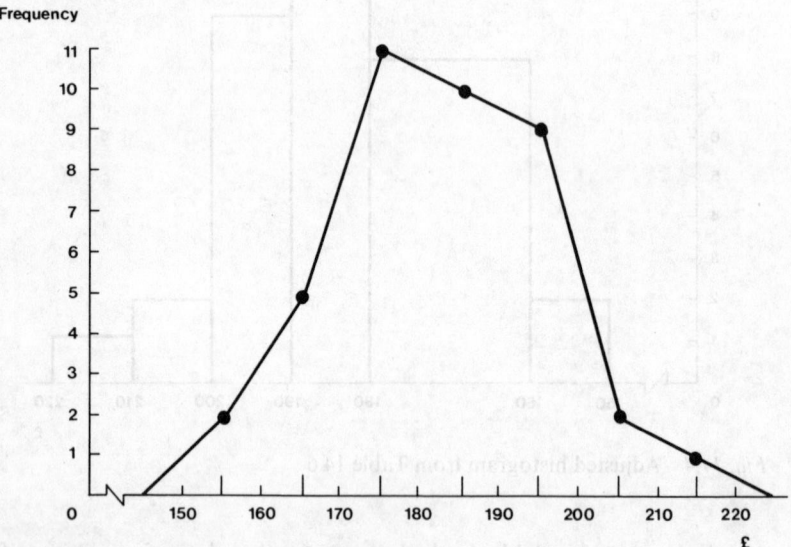

Fig. 14.5 Frequency polygon from Table 14.2

14.5.3 Frequency curves

The idea of representing frequencies by areas, which forms the basis of a histogram, can also be utilised if continuous curves are used to join the plotted frequency points.

Obviously the greater the number of classes represented by a histogram, the more closely the block diagram can approximate to a curve, but even with our data on rating valuations this idea can be shown.

Figure 14.6 shows a frequency curve superimposed on the histogram for these data. We are able to say from this diagram that, for instance, the number of valuations less than £160 is approximately represented by the shaded area under the curve.

The importance of this use of frequency curves will become apparent in further statistical work in ensuing chapters.

Graphical presentations of data

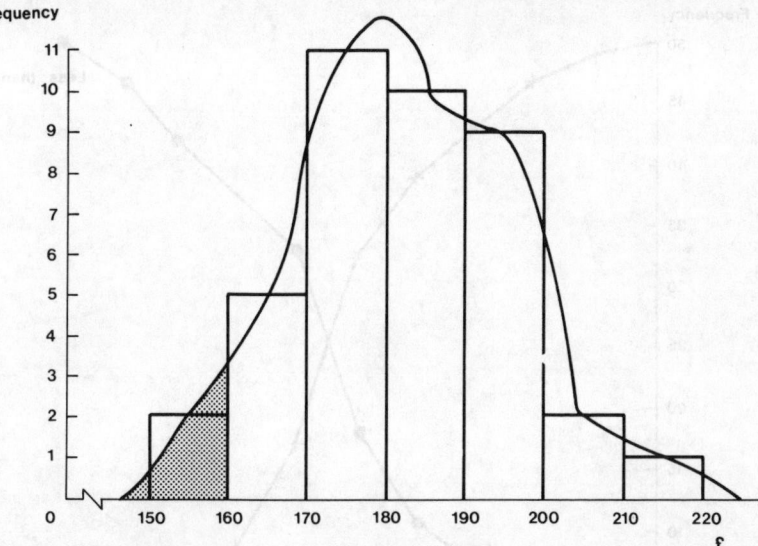

Fig. 14.6 Frequency curve and histogram from Table 14.2

14.5.4 Ogives

So far we have been concerned with drawing diagrams to represent frequency distributions. By applying the method of constructing a frequency polygon to a cumulative distribution we can draw a polygon called an ogive – so called from a term used in architecture for the characteristic shape formed by this cumulative frequency polygon.

Figure 14.7 shows ogives used to depict the cumulative frequency distributions in Tables 14.4 and 14.5. Here the cumulative frequencies are plotted instead of the ordinary frequencies.

Unlike the frequency polygon, though, the points are not plotted against the class mark, ie the middle of the class interval. In the 'less than' ogive, the cumulative frequencies are plotted against the upper class boundaries of 9.5, 19.5, etc. This is to differentiate this ogive from an 'or less' ogive.

There is obviously a difference between 'less than 10' and '10 or less' and we need to show this.

In the 'or more' ogive, however, the cumulative frequencies can be plotted against the 10, 20, 30, etc marks.

The ogive is often used for calculation problems, as we shall see in the next chapter, and so we usually join the points together by straight lines and not curves on the assumption that the items in each group are evenly spread between the boundaries of each class interval.

233

Data and its presentation

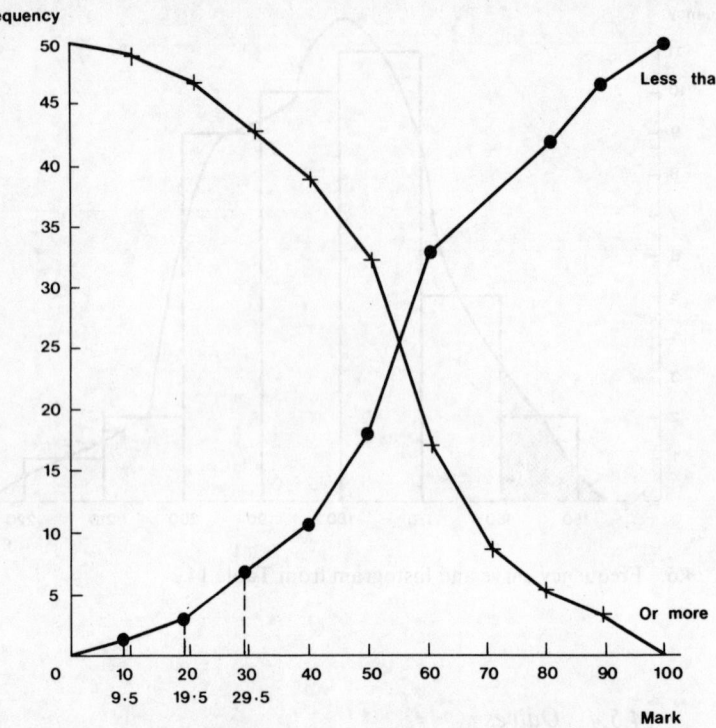

Fig. 14.7 Cumulative frequency curves (ogives) from Tables 14.4 and 14.5

14.6 OTHER METHODS OF DISPLAYING DATA

So far we have only been concerned with diagrams which are used to display data from frequency tables. There are many other methods of pictorial display of data and some of these are considered below.

14.6.1 Bar charts

A bar chart consists of bars of different heights drawn to depict the total number of items in a group. The blocks are always of equal width and usually differ from histogram blocks in that they do not normally have a scale measured along their base axis. (Note: A horizontal bar chart may sometimes be used instead of a vertical one.)

Example 14.1

Table 14.7 shows the numbers of houses of various types sold by two estate agents in a particular month.

A bar chart displaying the data in this table is shown in Fig. 14.8. Note that to make for easy understanding of the diagram, shading of the bars

Other methods of displaying data

Table 14.7 Number of residential properties sold

Estate agent	A	B
Town houses	8	6
Semi-detached	12	8
Detached	5	6
Total	25	20

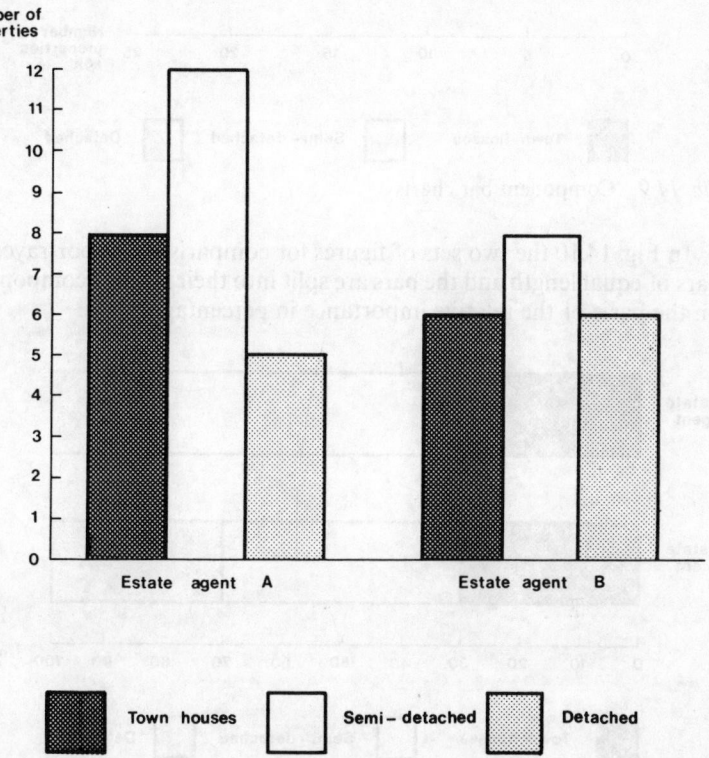

Fig. 14.8 Bar charts from Table 14.7

and a key have been used and also the same order of the different categories has been used.

To make it easier to compare the overall total sales of the two estate agents, while still being able to see the breakdown into the different categories of property, component bar charts may be used instead as in Fig. 14.9.

Another variation of the bar chart to consider is the percentage component bar chart. This is mainly used when it is important to view the relative importance of the various components.

235

Data and its presentation

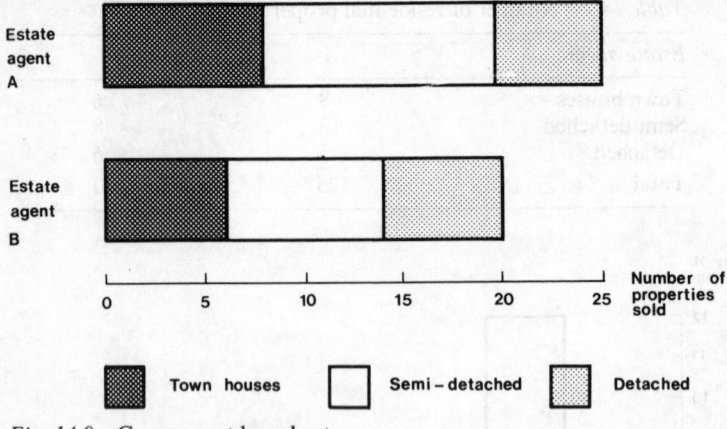

Fig. 14.9 Component bar charts

In Fig. 14.10 the two sets of figures for comparison are portrayed by bars of equal length and the bars are split into their various components on the basis of the relative importance in percentage terms.

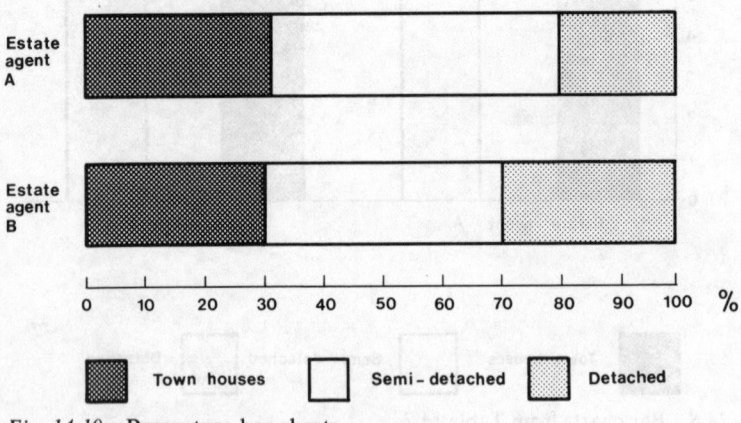

Fig. 14.10 Percentage bar charts

The percentage values are shown in Table 14.8.

Table 14.8 Percentage value of houses sold

	Estage agent A	Estate agent B
Town houses	8/25 × 100 = 32%	6/20 × 100 = 30%
Semi-detached	12/25 × 100 = 48%	8/20 × 100 = 40%
Detached	5/25 × 100 = 20%	6/20 × 100 = 30%

Other methods of displaying data

14.6.2 Pie charts

These diagrams are commonly used to represent data where ratios or relative values are important.

A pie chart is a circle divided into segments, the sizes of which are proportionate to the sizes of the components they represent.

To construct a pie chart the percentage of the total attributable to each component needs to be found and each percentage then multiplied by $\frac{360}{100}$ to obtain the number of degrees to assign to each component.

We have already calculated the percentage of sales made up by each category of property for estate agent A in Table 14.8. Multiplying each percentage by $\frac{360}{100}$, we obtain the following:

Town houses $\quad 32\% \times \frac{360}{100} = 115.2°$
Semi-detached $\quad 48\% \times \frac{360}{100} = 172.8°$
Detached $\quad 20\% \times \frac{360}{100} = 72°$

The pie chart of this data is shown in Fig. 14.11. Pie charts have a major advantage in that they show how one whole set of data is divided into various components and it is easy to see what size these components are in relation to each other and to the whole set.

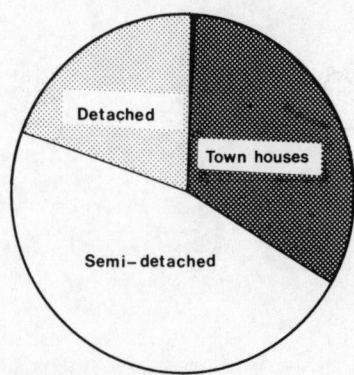

Fig. 14.11 Pie chart

14.6.3 Pictograms

Finally, one of the most common and striking ways of illustrating statistics is by means of a graph employing forms of pictures.

Small pictures of a particular object to portray the object involved are used, eg houses for construction data, coins for expenditure, men for employment, etc.

Examples of this type of diagram are in frequent use in the media, so just one case is presented here.

Table 14.9 shows the number of council houses sold in this country for a three-year period.

Data and its presentation

Table 14.9 Council house sales 1974–76 (*Housing and Construction Statistics*)

Year	Sales
1974	4557
1975	2723
1976	5793

A pictogram whereby one small house represents 1000 from the table, is shown in Fig. 14.12.

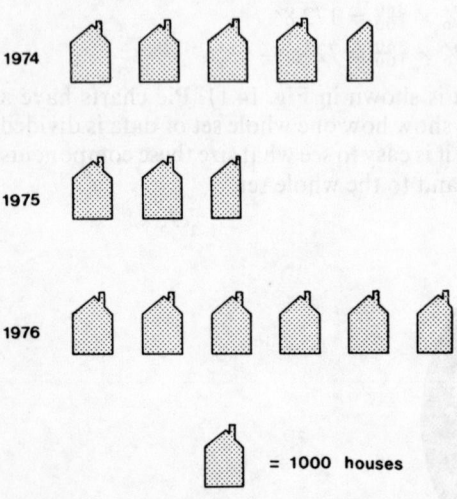

Fig. 14.12 Pictogram

Pictograms certainly have an advantage over the other types of diagram when it comes to attracting attention to the main features of the data portrayed, but cannot really play a part in any detailed statistical work because of their lack of precision.

Only a few of the more common types of diagram have been considered in this section. There are many different variations of the bar charts, pie diagrams and pictograms and really a great deal of flexibility is permissible in the drawing up of these diagrams. The main concern is that a diagram conveys its information in a concise and interesting manner.

Many examples of these various diagrams can be seen in trade journals, newspapers, etc and it should be an interesting exercise to the reader to look at these diagrams in practice.

14.7 EXERCISES

1. The following data gives the prices in £, of 50 houses in the same area.

8 800	10 100	8 025	9 300	9 575
9 700	8 750	10 200	8 375	11 800
9 550	8 700	8 450	9 200	9 200
8 525	9 450	8 475	8 800	7 825
8 700	7 650	6 775	9 225	9 400
9 750	6 100	8 300	8 675	8 200
7 250	9 100	7 800	9 300	8 150
10 700	10 850	8 900	10 900	8 700
9 900	11 200	9 650	6 650	9 800
10 400	8 150	8 950	8 875	9 250

Choose suitable class limits and put the above data into a frequency table.

Draw a histogram to illustrate the frequency table. (RICS)

2. The following marks were obtained by 50 building students in an examination.

94	40	70	57	48
56	75	67	62	67
25	53	52	43	55
45	49	55	45	53
58	44	74	63	42
50	65	59	46	36
59	25	77	54	23
33	83	47	43	52
29	47	61	50	37
79	41	51	63	49

(a) tabulate the results in the form of a frequency distribution;
(b) plot the result of (a) in the form of a histogram;
(c) construct a cumulative frequency table and plot the ogive from this table.

3. It is said that 'one picture is worth a thousand words'. How far might it also be correctly said that a picture is worth a thousand numbers?

Discuss this in relation to the use of pictorial and other diagrammatic methods in presenting various types of statistical data for various purposes to differing sorts of reader. (RICS)

15 MEASURES OF LOCATION

15.1 INTRODUCTION

So far two basic methods of dealing with sets of figures have been considered. Statistics have been put into tabulated form and into diagrammatic form. This type of exercise on its own, though, does not tell us very much about the characteristics of a group of numbers.

15.2 MEASURES OF CENTRAL TENDENCY

A measure of central tendency (or average) represents an attempt to produce one value from a whole set of data which can be regarded as representative of all the items.

There are several types of average employed in statistical analysis, but we shall concentrate on three of the more important ones.

1. The arithmetic mean.
2. The median.
3. The mode.

15.3 THE ARITHMETIC MEAN

This is what the layman would normally calculate if asked to find the average of a group of numbers, and is hereinafter referred to as just the mean. It is found by dividing the sum of the values by the number of items (or total frequency).

Example 15.1

The number of properties sold by an estate agency over an eleven-week period were:

6 8 8 4 3 10 8 1 4 1 2

$$\text{The mean} = \frac{\text{Sum of values}}{\text{Number of weeks}} = \frac{55}{11} = 5 \text{ properties}$$

15.3.1 Finding the mean from a frequency table

Ungrouped data. We can return to the example in the previous chapter (section 14.2) where the data on concrete building blocks were put into a frequency table.

It is possible to find the mean from the tabulated data by simply multiplying each value (mm) in the table by the corresponding frequency, summing the results and dividing the total by the total number of blocks.

It is worthwhile at this point introducing a symbolic notation into the calculations we are making.

As previously we can denote the frequency by the letter f, and we shall denote the column of values in the table by the letter x. Each value multiplied by its corresponding frequency is therefore fx.

When adding together all these fx products we write the total sum as Σfx, as the Greek capital letter Σ is a short-hand way of writing 'the sum of'. The total number of values we can therefore write as Σf, ie the sum of the frequencies.

So our formula for the mean of tabulated data can be written:

$$\bar{x} = \frac{\Sigma fx}{\Sigma f}$$

where $\bar{x}$ (pronounced x-bar) signifies the mean of the x's.
Table 15.1 shows the tabulated data on brick thicknesses.

$$\bar{x} = \frac{\Sigma fx}{\Sigma f} = \frac{2504}{25} = 100.16 \text{ mm}$$

Table 15.1 Frequency distribution of brick thicknesses

Brick thickness (mm) x	f	fx
97	1	97
98	3	294
99	4	396
100	8	800
101	3	303
102	4	408
103	2	206
	$\Sigma f = 25$	$\Sigma fx = 2504$

Measures of location

Grouped data. When using a similar procedure to find the mean of the grouped data a specific assumption is made concerning the values which fall in any particular class.

If we take as an example just one class from the frequency table of surveying examination marks in Table 14.3:

Marks f
20–29   4

We do not know without referring back to the original raw data what the exact marks of these four surveying students were – this loss of information being inevitable in the classification process. What we do therefore is to relate the frequency to the midpoint or class mark of each class.

The midpoint of a class is found by adding together the first and last possible values in a class and dividing by two. So here the midpoint is

$$\frac{20-29}{2} = 24.5$$

When finding the mean what we are basically assuming is that this value of 24.5 represents the whole class 20–29.

24.5 and the other midpoints from the frequency table are used as the x values to which the frequencies are applied.

We can now use Table 15.2, which shows the classified surveying marks to find the mean.

$$\bar{x} = \frac{\Sigma fx}{\Sigma f} = \frac{2655}{50} = 53.1 \text{ marks}$$

Table 15.2 Classified surveying marks

Marks	Midpoint x	f	fx
0–9	4.5	1	4.5
10–19	14.5	2	29.0
20–29	24.5	4	98.0
30–39	34.5	4	138.0
40–49	44.5	7	311.5
50–59	54.5	15	817.5
60–69	64.5	9	580.5
70–79	74.5	3	223.5
80–89	84.5	2	169.0
90–99	94.5	3	283.5
		$\Sigma f = 50$	$\Sigma fx = 2655.0$

The arithmetic mean

15.3.2 The assumed-mean method

The calculation involved in finding the mean from classified data can often be simplified by working from an assumed mean, which is then adjusted appropriately.

If we take the classified data on rating valuations from the previous chapter (as shown in Table 14.2) we could find the mean by firstly guessing at what the mean will be by inspection. We will take £184.5 as our estimate (note that this value is a midpoint of one of the classes; it simplifies matters if a midpoint value is chosen as the assumed mean).

The procedure is then to find by how much the other midpoints deviate from the assumed mean, to multiply each deviation (d) by the appropriate frequency and to sum these fd products.

The mean of the deviations ($=\Sigma df/\Sigma f$) is then used to adjust the assumed mean to the true mean. This procedure is illustrated in Table 15.3.

Table 15.3 Procedure to find assumed mean

Rating valuation (£)	Midpoint	Deviation from assumed mean of 184.5 (d)	f	df
150–159	154.5	−30	2	−60
160–169	164.5	−20	5	−100
170–179	174.5	−10	11	−110
180–189	184.5	0	10	0
190–199	194.5	10	9	90
200–209	204.5	20	2	40
210–219	214.5	30	1	30
			$\Sigma f = 40$	$\Sigma df = -110$

The mean of the deviations $= \dfrac{\Sigma df}{\Sigma f} = \dfrac{-110}{40} = -2.75$

∴ The true mean = assumed mean + the mean of the deviations
$$= 184.5 + (-2.75)$$
$$= £181.75$$

15.3.3 Weighted means

In some situations it is unsatisfactory to calculate the mean of a set of items without taking account of the relative importance of certain items. In such circumstances it is appropriate to calculate a weighted mean.

Measures of location

The following example shows the use of a weighted mean. A builder's merchant sells three types of screen wall blocks (types A, B and C) and he makes 10 pence, 20 pence and 30 pence profit respectively on each unit that he sells. His records show that for every 5 blocks of type A that he sells, he is able to sell 3 blocks of type B and 2 blocks of type C.

If we want to find the merchant's mean profit per unit sold, we should not merely add together 10, 20 and 30 and divide by 3 to give a mean of 20 pence. Instead the three profit figures should be weighted according to how important each type of block is in the merchant's sales.

We find a weighted mean by attaching weights to each of the profit figures and then dividing by the total weights. The general formula is:

$$\text{Weighted mean } (\bar{x}_w) = \frac{w_1 x_1 + w_2 x_2 + \cdots + w_n x_n}{w_1 + w_2 + \cdots + w_n}$$

where

$x_1, x_2, \ldots x_n$ = then n items for which we wish to find the mean

$w_1, w_2, \ldots w_n$ = the weights attached to each item $x_1, x_2, \ldots x_n$ respectively.

So in our example the x's are the profit figures and the w's are taken from the sales figures. We have then:

$x_1 = 10 \quad x_2 = 20 \quad x_3 = 30$

$w_1 = 5 \quad w_2 = 3 \quad w_3 = 2$

Putting these figures into the formula gives:

$$\bar{x}_w = \frac{10(5) + 20(3) + 30(2)}{5 + 3 + 2}$$

$$= \frac{170}{10} = 17 \text{ pence}$$

As we shall see in Ch. 18, this type of 'averaging' using weights is often a necessary procedure in the composition of index numbers.

15.4 THE MEDIAN

When a number of values are arranged in numerical (ascending or descending) order, the median is the middle value.

If the figures on property sales from Example 15.1 are put into ascending order we have:

1 1 2 3 4 4 6 8 8 8 10

The middle value is the 6th one and is 4. The simple rule is to take the value of the $((n + 1)/2)$th item in a series, where n is the total number of items, so in this example because we have an odd number of values, we had an exact item in the list to refer to.

If we had only ten figures in our original list, eg if the last figure (2) had been omitted, the ordered figures would have been:

1 1 3 4 4 6 8 8 8 10

and the median value is the $((10 + 1)/2)$th in the list which means it is midway between the 5th and 6th values. The median is therefore, $(4 + 6)/2 = 5$

15.4.1 Finding the median from a frequency table

Ungrouped data. Once data have been put into frequency table form, they have already been put into numerical order so it is a relatively easy matter to read off the appropriate value from the table.

In the frequency table on brick thickness (Table 15.1) the total frequency is 25 and so the median value is that of the $((25 + 1)/2)$th = 13th. To find out what this value is we can refer to the appropriate cumulative frequency table (Table 15.4). From the table it can be seen that there are 8 bricks with thicknesses of 99 mm or less, and 16 with thicknesses of 100 mm or less.

The middle one (the 13th) must therefore be one of those with a thickness of 100 mm.

Table 15.4 Cumulative frequency table for brick thicknesses

Brick thickness (mm)	Cumulative frequency
97 or less	1
98 or less	4
99 or less	8
100 or less	16 ←
101 or less	19
102 or less	23
103 or less	25

Grouped data. From classified frequency tables we can find the class containing the median value in a manner similar to that just described, but then have the additional problem of locating the median within the appropriate class. The data on rating valuations are illustrated in Table 15.5.

The median of a set of grouped data is worked out according to a slightly different formula than for ungrouped data. The median is the value of the n/2th item. So here we are interested in the rating valuations of the 20th property.

Measures of location

Table 15.5 Rating valuations

Rating valuation assessment (£)	Number of properties	Cumulative frequency
Less than 160	2	2
160 but less than 170	5	7
170 but less than 180	11	18
180 but less than 190	10	28
190 but less than 200	9	37
200 but less than 210	2	39
210 but less than 220	1	40

It can be seen from the table that this property has a rating valuation somewhere between £180 and £189. We now make an assumption that the valuations of the 10 properties in this class are evenly distributed between the two class limits. The 20th property overall is the 2nd out of the 10 in this class and so we assume that this valuation is 2/10th of the way along this class interval.

The class interval here is from the lower class boundary of the £179.5 to the upper class boundary of £189.5 and is therefore equivalent to £10.

The 20th property is assumed to be located at a point 2/10 of the way along this interval of 10 which begins at £179.5, ie

$$\text{the median valuation} = 179.5 + (\tfrac{2}{10} \times 10)$$
$$= \underline{£181.5}$$

This general formula for finding the median for grouped data can be presented more formally as:

$$\text{Median} = L_m + \left(\frac{\tfrac{n}{2} - F_{m-1}}{f_m} \right) \times C$$

where

L_m = the lower class boundary of the median group.
C = the class interval.
F_{m-1} = the cumulative frequency for the preceding group.
f_m = the frequency of the median group.

15.4.2 Finding the median from an ogive

The median can be found graphically quite simply from an ogive.

From the plotted curve the value of the middle item from the cumulative frequency scale is read off. Figure 15.1 shows the 'less than' ogive for the rating valuation data in Table 15.5. We can read off from the ogive the value corresponding to the 20th property as £181.5.

The mode

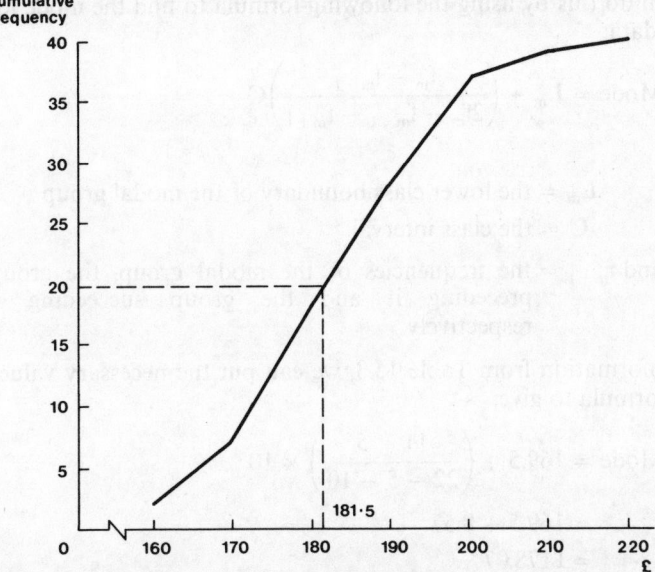

Fig. 15.1 Use of the ogive to find the median

15.5 THE MODE

The mode may be defined as the value which occurs most often.

Whenever we have unclassified data, there is no calculation required at all for the mode. It is merely found by the selection of the value with the highest frequency.

From the data in Example 15.1, the value which occurs most commonly is 8, ie on three occasions this number of properties were sold and no other value occurs three times.

To find the mode from a frequency table such as Table 15.1 is a simple matter. The value 100 mm has a frequency of 8 and is the most common value.

With a classified frequency table it is easy to select the modal class and often that is all that we are concerned with. Difficulties arise when we wish to decide exactly where within the modal group, the mode lies.

We could assume that the items in the class are evenly distributed and choose the midpoint as our estimate, but if we look at the frequency table on rating valuations in Table 15.3 the justification of doing this might be questioned. The modal class is 170 to 179 but there are twice as many valuations between 180 and 189 as between 160 and 169.

It is therefore quite valid to consider this lack of symmetry in the distribution of the valuations by taking account of the frequencies of the classes preceding and succeeding the modal class.

Measures of location

We can do this by using the following formula to find the mode for grouped data:

$$\text{Mode} = L_m + \left(\frac{f_m - f_{m-1}}{2f_m - f_{m-1} - f_{m+1}}\right)C$$

where

L_m = the lower class boundary of the modal group
C = the class interval.
f_m, f_{m-1} and f_{m+1} = the frequencies of the modal group, the group preceding it and the group succeeding it respectively.

Taking information from Table 15.3, we can put the necessary values into the formula to give:

$$\text{Mode} = 169.5 + \left(\frac{11 - 5}{22 - 5 - 10}\right) \times 10$$

$$= 169.5 + 8.57$$

$$= \underline{£178.07}$$

As f_{m+1} was much greater than f_{m-1}, here, the modal value is deemed to be closer to the upper class boundary than the lower one.

15.6 OTHER AVERAGES

There are two other types of mean, which are, in certain specific situations, appropriate types of average. These are:

The geometric mean. This average is mainly used in situations where relative values are compared in a ratio form particularly with index numbers.

The geometric mean of n numbers is the nth root of their product, eg the geometric mean of 1, 2, and 3 is $\sqrt[3]{1 \times 2 \times 3} = 1.82$.

The geometric mean is less than the arithmetic mean in this example and the fact that it is not as affected by abnormal values is one reason why it is sometimes preferred to the arithmetic mean.

The harmonic mean. This average can be described as the reciprocal of the arithmetic mean of the reciprocals of the items.

The harmonic mean of 1, 2, and 3 is:

$$\frac{3}{\frac{1}{1} + \frac{1}{2} + \frac{1}{3}} = 1.64$$

This type of mean is only applicable in very special cases, and is only included in this section to complete the list of averages.

15.7 A COMPARISON OF THE MEAN, MEDIAN AND MODE

The main purpose of all these types of average is to describe the centre of a set of data, but each of them performs this function in a different, specific way.

The question of which average should be used in a given situation is not an easy one to answer.

Generally speaking, the arithmetic mean is the most important of these measures because it is the most commonly understood average and also because of its suitability for further statistical work. To calculate the mean all the data must be utilised; whereas this is not the case with the other two measures.

The main merit of the median is that it is an average not affected by high or low value items; whereas the mean can give a distorted view of a set of data where even just one or two items lie outside the main body of values.

If we look at a simple example the following figures show the number of storeys of ten commercial buildings in a particular area.

2 2 2 3 3 3 3 3 9 10

The mean height = $\frac{40}{10}$ = 4 storeys.

It is fairly obvious that this is not a 'good' average to choose in this situation as it could not reasonably be judged to be representative of the data. There is not a single building with that number of storeys. Eight out of ten buildings have less storeys than this and it is only the heights of the two untypically tall buildings which drag the mean value up to that level. The median height of three storeys on the other hand is much more representative of the typical building height as it is unaffected by the two 'abnormal' values.

The mode has the major advantage of being an actual value in a set of data, whereas the mean value may not, but is less useful as an average when a set of data is widely dispersed with no obvious modal point or with more than one modal point.

Overall the arithmetic mean is the most useful statistic to represent central tendency.

15.8 QUARTILES, DECILES AND PERCENTILES

So far we have considered only measures of central location, ie measures which give us information on the centre of a set of data.

The median value is that which divides a group of items into two equal parts. In many circumstances, though, we may be more concerned with finding such things as the value above which three quarters of the items lie or the value below which 90 per cent of the values can be found, etc.

Measures of location

Quartiles. If we wish to divide a set of data into four equal parts, then we find the quartile values.

The lower quartile (Q_1) is the value below which one quarter of the items lie and the upper quartile (Q_3) divides the data such that three quarters of the items fall below it and one quarter above.

The remaining quartile (Q_2) does not have to be separately defined because it is equivalent to the median.

Deciles and percentiles. Deciles are defined as those values which divide a set of items into ten equal parts. Similarly the percentile points divide the data into a hundred equal parts.

These measures are frequently used whenever it is required that a set of data needs to be divided up into certain fixed proportions. For instance, in the example on surveying examination marks the pass mark may be set according to the requirement that 60 per cent of the students pass the examination and 40 per cent fail. The fortieth percentile (fourth decile) therefore needs to be found.

The procedure for finding any of these above measures of location is exactly the same as the one explained for finding the median.

15.9 EXERCISES

1. In a painting test the area covered by 500 ml of primer was measured in square metres. The results of fifty such tests were

Area (mid-point)	4.6	4.7	4.8	4.9	5.0	5.1	Total
Frequency	2	7	17	15	6	3	50

   Calculate the arithmetic mean of the above data. Find two other averages for the data either by calculation or graphically. Discuss the advantages and disadvantages of these averages, then state, with reasons, which gives the best one for the data. (RICS)

2. The earnings of seventy-two workers on a building site are shown below:

Wage (£)	Number of workers
81–90	12
91–100	14
101–110	22
111–120	17
121–130	4
131–140	2
141–150	1

Using graphical methods estimate the median wage. Comment on the use of the median to interpret this data and the reasons why it may be preferable to the mean in this case.

3. Find the arithmetic mean of the following weekly wage distribution:

Wage (£)	35	40	45	50	55	60	65	70	75	80	85
Frequency	3	20	40	30	20	10	9	6	5	4	3

Draw a histogram of the above data and use the diagram to determine the value of the mode.

Describe another type of average that could be found from the data.

Discuss which would be the most appropriate average to use for the data above. (RICS)

16 MEASURES OF DISPERSION

16.1 INTRODUCTION

The different averages dealt with so far sum up certain characteristics of a set of data. An average value is a typical value and is often the most important statistic used to describe data.

To give a more complete description of the data we are dealing with, though, we need to consider the spread or dispersion of the data.

To show the importance of finding a measure of the dispersion of the items in a set of data, refer to Table 16.1, which shows the numbers of bricks laid by two teams of bricklayers during a certain time period.

Table 16.1 Bricks laid by two teams

Worker number	Team A Number of bricks laid	Team B Number of bricks laid
1	80	96
2	91	99
3	98	100
4	111	102
5	120	103

For both teams the mean number of bricks laid is 100 but the groups are far from identical, the first group of numbers having a much greater spread than the second.

In situations such as this we need some statistic which will indicate such differences.

The main measures of dispersion are:
(a) the range;
(b) the quartile deviation;
(c) the mean deviation;
(d) the standard deviation.

We can look at each of these in turn.

16.2 THE RANGE

This is the most straightforward measure of dispersion. We calculate the range simply by subtracting the lowest item from the highest.

In our example the range from Team A is $120 - 80 = 40$ and for Team B it is $103 - 96 = 7$. If this statistic were given along with the mean, it would indicate the obvious difference between the two sets of data. The range is the easiest measure of dispersion to understand and the simplest to obtain, but as it ignores the majority of data and does not tell us anything about the dispersion between the lowest and highest values, it is considered to be a fairly crude statistic.

16.3 THE QUARTILE DEVIATION

The quartiles may be employed to arrive at a measure of dispersion. This measure is known as the quartile deviation and is obtained by halving the inter-quartile range, ie the difference between the lower and upper quartiles.

It can be written as:

$$\text{Quartile deviation} = \frac{Q_3 - Q_1}{2}$$

Whilst it is true that the quartile deviation is not affected by extreme variations – a disadvantage of the range as a measure of dispersion – the fact that the first and last quarters of the data are ignored means that it is not a satisfactory measure of the dispersion of the whole data.

As it is important to know the spread of all the items in the data, other measures are therefore used.

16.4 THE MEAN DEVIATION

This is a measure which makes use of all the data and is a statistically more satisfactory method than the range and quartile deviation.

The mean deviation gives the mean of the deviations of items from the mean, median or mode. (It is usually more important to calculate the

Measures of dispersion

mean deviation of the items from the mean because this is the 'average' most useful for further statistical work.)

The reason why we are interested in the mean deviation from the mean is that the dispersion of a set of data will be small if the items deviate in value from the mean by only a small amount, but the deviations will be large if the items are widely dispersed in value.

The calculation of the mean deviation (from the mean) is illustrated in Table 16.2, which relates to the data on students' surveying marks. The mean mark is 53.1.

Table 16.2 Calculation of mean deviation

Mark	Midpoint	f	Deviation of midpoint from mean of 53.1 $(x - \bar{x})$	$f(x - \bar{x})$
0–9	4.5	1	48.6	48.6
10–19	14.5	2	38.6	77.2
20–29	24.5	4	28.6	114.4
30–39	34.5	4	18.6	74.4
40–49	44.5	7	8.6	60.2
50–59	54.5	15	1.4	21.0
60–69	64.5	9	11.4	102.6
70–79	74.5	3	21.4	64.2
80–89	84.5	2	31.4	62.8
90–99	94.5	3	41.4	124.2
		$\Sigma f = 50$		$\Sigma f(x - \bar{x}) = 749.6$

We are interested in the deviations of the marks from this mean and so the fourth column shows the deviation of each midpoint from 53.1 and this is shown as $(x - \bar{x})$. In the final column these deviations are multiplied by the frequency associated with each class and the products then summed.

So we have the sum of the deviations:

$$\Sigma f(x - \bar{x}) = 749.6$$

To find the mean of the deviations we divide by $\Sigma f (= n)$ to give:

$$\text{Mean deviation} = \frac{\Sigma f(x - \bar{x})}{\Sigma f} = \frac{749.6}{50} = 14.99 \text{ marks}$$

In this calculation the signs of the deviation (+ or −) were ignored, ie negative signs were not used when the midpoint value was less than the mean value. The reason for this is obvious when the last column of Table 16.2 is considered.

If the first five values in the deviation column had their negative signs inserted, the first five values in the $f(x - \bar{x})$ column would have totalled -374.8. This would have exactly matched the positive total of the last five values in the $f(x - \bar{x})$ column.

The deviations would obviously always total zero if the signs were included. The fact that the signs of the deviations are discarded is one of the reasons why the mean deviation is considered to be rather an artificial measure and unsuitable for further statistical work.

In practice, the most useful measure of dispersion for statistical work is the standard deviation which involves an approach similar to that of the mean deviation but does not include the drawbacks associated with the latter measure.

16.5 THE STANDARD DEVIATION

16.5.1 Introduction

This measure, which is by far the most important measure of dispersion, is basically a modification of the calculation of the mean deviation which incorporates two extra steps.

Instead of ignoring the signs of the deviations, the deviations are squared, which removes that problem. The squared deviations are added together and the total is divided by the number of items. The result is termed the **variance**.

The square root of the variance is the standard deviation.

This can be written as:

$$\text{Standard deviation} = \sqrt{\left(\frac{\text{Sum of squares of deviations from the mean}}{\text{Number of items}}\right)}$$

A simple illustration on the calculation of the standard deviations can be given if we consider the earlier example on the number of bricks laid by the first team of bricklayers. The method is shown in Table 16.3.

Table 16.3 Calculation of standard deviation

Worker number	Team A Number of bricks laid	Deviation from mean of 100 $(x - \bar{x})$	$(x - \bar{x})^2$
1	80	-20	400
2	91	-9	81
3	98	-2	4
4	111	11	121
5	120	20	400
		$\Sigma(x - \bar{x})^2 =$	1006

Measures of dispersion

We have already found the mean to be 100.
The total sum of the squared deviations is 1006 and so the variance is given by this total divided by the number of workers

$$\text{Variance} = \frac{\Sigma(x - \bar{x})^2}{n} = \frac{1006}{5} = 201.2$$

$$\text{The standard deviation} = \sqrt{\left(\frac{\Sigma(x - \bar{x})^2}{n}\right)} = \sqrt{(201.2)} = 14.18$$

16.5.2 Calculating the standard deviation from a frequency distribution

To find the standard deviation of data in a frequency table we use the same procedure as in the previous example but we need to take account of the frequencies.

The formula for finding the standard deviation from a frequency table is:

$$\text{Standard deviation} = \sqrt{\left(\frac{\Sigma f(x - \bar{x})^2}{\Sigma f}\right)}$$

Table 16.4 shows how the standard deviation can be calculated from the frequency table of students' surveying marks.

Inserting the necessary values for the formula, we obtain:

$$\text{Standard deviation} = \sqrt{\left(\frac{\Sigma f(x - \bar{x})^2}{\Sigma f}\right)} = \sqrt{\left(\frac{20\,202}{50}\right)}$$

$$= \sqrt{(404.04)} = 20.1 \text{ marks}$$

Table 16.4 Calculation of standard deviation from frequency table

Mark	Midpoint	f	$(x - \bar{x})$	$(x - \bar{x})^2$	$f(x - \bar{x})^2$
0–9	4.5	1	−48.6	2361.96	2361.96
10–19	14.5	2	−38.6	1489.96	2979.92
20–29	24.5	4	−28.6	817.96	3271.84
30–39	34.5	4	−18.6	345.96	1383.84
40–49	44.5	7	−8.6	73.96	517.72
50–59	54.5	15	1.4	1.96	29.40
60–69	64.5	9	11.4	129.96	1169.64
70–79	74.5	3	21.4	457.96	1373.88
80–89	84.5	2	31.4	985.96	1971.92
90–99	94.5	3	41.4	1713.96	5141.88
		$\Sigma f = 50$			20 202.00

16.5.3 Calculation of the standard deviation by the assumed mean method

The calculation involved in finding the standard deviation from a frequency table can be quite complicated even with the use of a calculator.

To simplify the arithmetical work a method based on the use of an assumed mean can be employed. This method is also suitable in a situation where it is desired to find the standard deviation of a set of data but not necessary to find the mean to start with.

In the surveying marks example we might take an assumed mean of 54.5. We could then use the formula for the standard deviation taking deviations from 54.5 and make an adjustment for the difference between the assumed mean and the true mean.

As we saw in section 15.3.2 the mean of the deviations of the x values ($\Sigma fd/\Sigma f$) was used to adjust the assumed mean to the true mean when we were using this method to find the mean.

We can incorporate this into our standard deviation formula by using the square of this adjustment.

This means that the formula for the standard deviation using this method is:

$$\text{Standard deviation} = \sqrt{\left[\frac{\Sigma d^2 f}{\Sigma f} - \left(\frac{\Sigma df}{\Sigma f}\right)^2\right]}$$

where d = the deviation of each x value from the assumed mean.

The method can be illustrated by Table 16.5, where another simplification is made as the deviation figures are divided by 10 initially in order to reduce the arithmetic.

Table 16.5 Calculation of standard deviation using assumed mean

Mark	Midpoint	f	Deviation from assumed mean of 54.5 in 10s d	df	d^2	$d^2 f$
0–9	4.5	1	−5	−5	25	25
10–19	14.5	2	−4	−8	16	32
20–29	24.5	4	−3	−12	9	36
30–39	34.5	4	−2	−8	4	16
40–49	44.5	7	−1	−7	1	7
50–59	54.5	15	0	0	0	0
60–69	64.5	9	1	9	1	9
70–79	74.5	3	2	6	4	12
80–89	84.5	2	3	6	9	18
90–99	94.5	3	4	12	16	48
				$\Sigma df = -7$		$\Sigma d^2 f = 203$

Measures of dispersion

The formula we need to use here is therefore

$$\text{Standard deviation} = C\sqrt{\left[\frac{\Sigma d^2 f}{\Sigma f} - \left(\frac{\Sigma df}{\Sigma f}\right)^2\right]}$$

where C = the class interval.

Putting the appropriate values into the formula:

$$\text{Standard deviation} = 10\sqrt{\left[\frac{203}{50} - \left(\frac{-7}{50}\right)^2\right]}$$

$$= 10 \times \sqrt{(4.06 - 0.0196)}$$
$$= 10\sqrt{(4.0404)}$$
$$= 20.1 \text{ marks}$$

So we have arrived at exactly the same answer as we did under the ordinary method, but have somewhat reduced the arithmetic involved.

16.5.4 A note on the formula for the standard deviation

The most important context in which a calculation of the standard deviation of a set of data is made occurs when an attempt is being made to make inferences about a population of items from a sample.

We can define a population as a complete set of individual objects or measurements, eg all the houses in a particular area, all the bricks produced by a manufacturer, etc. A sample, on the other hand, includes only some of the houses in an area or some of the bricks produced by a manufacturer.

How this distinction affects the formula for the standard deviation is that if we know that we are calculating the standard deviation from sample data, the denominator of the formula should be $n - 1$ (or $(\Sigma f) - 1$) replacing n (or Σf).

The formula for raw data would therefore become:

$$\text{Standard deviation} = \sqrt{\left(\frac{\Sigma(x - \bar{x})^2}{n - 1}\right)}$$

For data in a frequency table:

$$\text{Standard deviation} = \sqrt{\left(\frac{\Sigma f(x - \bar{x})^2}{(\Sigma f) - 1}\right)}$$

So in situations where it is indicated that the data is sample data, the above modified formulae should be used. Using $(n - 1)$ is important when n is small but if n is large (say more than 100) the difference between using $(n - 1)$ and n would be negligible.

16.6 THE COEFFICIENT OF VARIATION

One main purpose of finding a measure of dispersion for different sets of data is to enable comparisons of the variation in distributions to be made.

The standard deviation is measured in absolute terms but it can be converted to a relative form termed the coefficient of variation. This is defined as:

$$\text{Coefficient of variation} = \frac{\text{standard deviation}}{\text{mean}}$$

or commonly as:

$$\text{Coefficient of variation} = \frac{100 \times \text{standard deviation}}{\text{mean}}\%$$

The standard deviation is expressed either as a fraction or a percentage of the mean and thus becomes a relative measure of dispersion.

We can consider an example of two distributions with the same variable but different means and standard deviations.

Example 16.1

The mean price of houses sold in a particular year in area A was £21 000 with a standard deviation of £3325, whereas the mean price in B was £22 500 with a standard deviation of £3375.

We have

$$\text{Coefficient of variation (Area A)} = \frac{100 \times 3325}{21\,000}\% = 15.83\%$$

$$\text{Coefficient of variation (Area B)} = \frac{100 \times 3375}{22\,500}\% = 15.00\%$$

Thus property prices in Area A were relatively more variable than in Area B.

16.7 A NOTE ON THE INTERPRETATION OF THE STANDARD DEVIATION

A proper understanding of this most important measure can only really be gained by looking at its relationship to the normal distribution. We shall do this in the next chapter.

Measures of dispersion

16.8 SKEWNESS

Having dealt with the need to provide information on the dispersion of data in addition to an average value, there may be situations where to provide an adequate description of a set of data an additional type of statistic is required.

This additional statistic is a measure of the symmetry or skewness of a distribution.

It is possible that two distributions could have identical means and standard deviations yet differ greatly in their appearance.

The two distributions depicted by the frequency polygons in Fig. 16.1 both have a mean of 3.5 and a standard deviation of 1.2 yet the first one is perfectly symmetrical and the second one is skewed.

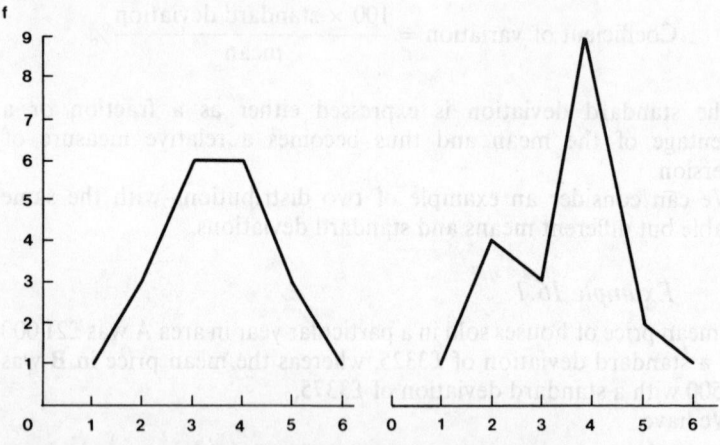

Fig. 16.1 (a) Symmetrical distribution (b) skew distribution

The way in which we can measure the degree of skewness of a distribution is to base the method on the fact that the greater the lack of symmetry in a distribution the greater will be the divergence between the mean, the median and the mode values. Fig. 16.2 illustrates this point.

Figure 16.2(a) This frequency curve is symmetrical and the mean, mode and median values are all exactly equal.

Figure 16.2(b) Here the frequency curve tails off to the right and the mean value is higher than the mode and median values. We say that the distribution is positively skewed.

Figure 16.2(c) The frequency curve tails off to the left and the mean value is less than the mode and median values. The distribution is negatively skewed.

A measure which shows the relative degree of skewness of a distribution is Pearson's coefficient of skewness.

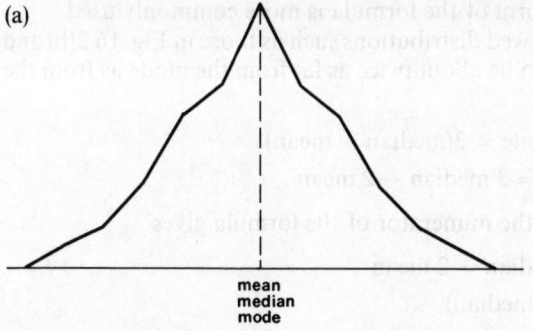

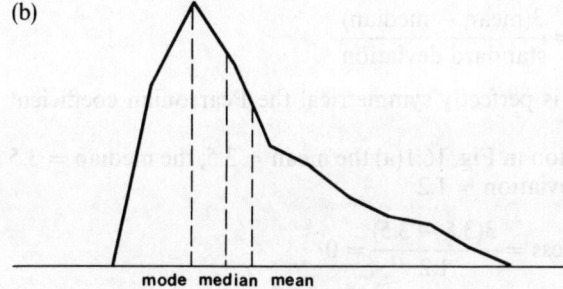

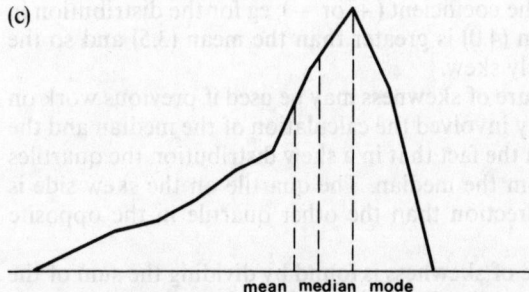

Fig. 16.2 (a) Symmetrical distribution; (b) positively skewed distribution; (c) negatively skewed distribution

The formula for this coefficient is:

$$\text{Skewness} = \frac{\text{mean} - \text{mode}}{\text{standard deviation}}$$

The main problem with using this formula is that it involves the mode, and as we saw in section 15.5, the mode is often difficult or impossible to find.

Measures of dispersion

So an alternative form of the formula is more commonly used.

For moderately skewed distributions such as those in Fig. 16.2(b) and (c), the median tends to be about twice as far from the mode as from the mean, ie

$$\text{median} - \text{mode} = 2(\text{median} - \text{mean})$$
$$\therefore \text{mode} = 3 \text{ median} - 2 \text{ mean}$$

Substituting this into the numerator of the formula gives

$$\text{mean} - 3 \text{ median} + 2 \text{ mean}$$
$$= 3(\text{mean} - \text{median})$$

Thus the alternative formula is

$$\text{Skewness} = \frac{3(\text{mean} - \text{median})}{\text{standard deviation}}$$

If a distribution is perfectly symmetrical the Pearsonian coefficient will be zero, eg

For the distribution in Fig. 16.1(a) the mean = 3.5, the median = 3.5 and the standard deviation = 1.2

$$\therefore \text{Skewness} = \frac{3(3.5 - 3.5)}{1.2} = 0$$

If a distribution is skew though, the direction of the skewness will determine the sign of the coefficient ($+$ or $-$), eg for the distribution in Fig. 16.1(b), the median (4.0) is greater than the mean (3.5) and so the distribution is negatively skew.

An alternative measure of skewness may be used if previous work on a set of data has already involved the calculation of the median and the quartiles. It is based on the fact that in a skew distribution the quartiles are not equidistant from the median. The quartile on the skew side is pulled more in that direction than the other quartile in the opposite direction.

The relative measure of skewness is found by dividing the sum of the quartiles less twice the median by the quartile deviation, ie

$$\text{Skewness} = \frac{Q_3 + Q_1 - 2 \text{ median}}{(Q_3 - Q_1)/2}$$
$$= \frac{2(Q_3 + Q_1 - 2 \text{ median})}{Q_3 - Q_1}$$

16.9 EXERCISES

1. 80 students sat examinations in building technology and in construction.

The marks were distributed as follows:

Marks	Number of students	
	Building technology	Construction
1–10	1	0
11–20	3	1
21–30	5	1
31–40	7	11
41–50	12	13
51–60	21	30
61–70	18	12
71–80	8	11
81–90	4	1
91–100	1	0

Calculate the mean and standard deviation of each frequency table and compare the spread of marks in each examination.

2. A branch of a building society has 900 customers who have bought property within certain price ranges. The frequency distribution of these prices is shown below.

Price range (£)	Number of customers
20 000–21 999	24
22 000–23 999	38
24 000–25 999	110
26 000–27 999	154
28 000–29 999	290
30 000–31 999	142
32 000–33 999	68
34 000–35 999	46
36 000–37 999	28

Find the median price and an appropriate measure of dispersion.

3. A survey was carried out to measure the distance travelled each day from home to work. The results, in kilometres, are tabulated below.

Distance (mid-point)	1	3	5	7	9	11	13	15	17	19
Frequency		2	4	9	18	21	19	17	7	2	1

Calculate the arithmetic mean and the standard deviation of the above data.

Use the data to draw the cumulative frequency graph and use this graph to find another average. Give both the name and the value of this average.

What percentage of the sample travel more than 15 km to work?

(RICS)

17 PROBABILITY AND THE NORMAL DISTRIBUTION

17.1 PROBABILITY

The term probability is a difficult one to define in simple terms. In statistical work we are mainly concerned with the problem of obtaining a numerical measure of probability and there are two ways of approaching this type of problem.

17.1.1 The theoretical approach

This can best be considered by reference to games of chance. If we look at the simple example of throwing an unbiased die, there are six possible outcomes, ie obtaining a one, two, three, four, five or six when the die is thrown and each outcome is equally likely to occur. So the probability of throwing a six, P(six)–or indeed any other number–can be reasoned to be $\frac{1}{6}$, which was deductively calculated by:

$$P(six) = \frac{\text{Number of outcomes involving a six}}{\text{Total number of possible outcomes}}$$

This probability of $\frac{1}{6}$ is sometimes termed the relative frequency of obtaining a six.

In an example where all outcomes have an equal chance of occurring, the probability of a particular event is the number of outcomes favourable (m) to that event divided by the total number of possible outcomes (n)

$$P(event) = \frac{m}{n}$$

We can also say that the probability that an event does not occur is $(n - m)/n$ and in our example

$$P(\text{not throwing a six}) = P(\text{one, two, three, four or five}) = \tfrac{5}{6}$$

The rules of probability

Thus total probability, ie the sum of the probabilities of all possible outcomes, equals one. This is termed *certainty*.

P(six) + P(not obtaining a six) = $\frac{1}{6} + \frac{5}{6} = 1$.

One of these two events is certain to occur.

1 is an extreme value of probability, the other extreme being 0. If P(event) = 0, the event is an *impossibility*.

17.1.2 The empirical approach

This involves a study of how often an event occurs in a large number of trials.

Expected relative frequencies can be applied in situations such as tossing a coin, throwing a die, etc, but not in many real life situations. It may be necessary to derive expected relative frequencies from empirical findings.

For example, a random sample of 100 students at a college may show that ten of the students have blond hair. So we could estimate that the proportion of all the students in the college with blond hair is 1 in 10 and thus state that the probability of any student chosen at random having blond hair is $\frac{1}{10}$.

17.2 THE RULES OF PROBABILITY

There are two basic rules which can be applied to probability. The first is the addition rule. If there are two or more ways in which an event can occur, then as long as the separate ways are *mutually exclusive* (ie two or more cannot occur simultaneously) the probability of its occurring is the sum of the probabilities of each of the different ways.

Thus the probability of obtaining a five or six when a die is thrown is $(\frac{1}{6} + \frac{1}{6}) = \frac{1}{3}$.

If we are concerned with the probability of two or more events occurring simultaneously or in succession, then the multiplication rule must be used.

The rule is that the probability of the occurrence of such events is the product of the separate probabilities of each event.

The probability of obtaining two sixes when two die are thrown is $(\frac{1}{6} \times \frac{1}{6}) = \frac{1}{36}$. The two events–obtaining a six on one die and obtaining a six on the second die–are termed *independent events*, because the occurrence of one event in no way affects the occurrence of the other.

If events are non-independent, adjustments to the probabilities used need to be made. If we are concerned with the probability of obtaining two aces when a pack of cards is cut twice, then the probability of choosing an ace initially is $\frac{4}{52}$ (or $\frac{1}{13}$) and if an ace is chosen and then returned to the pack and a choice from fifty two cards is again made the probability of a second ace is also $\frac{1}{13}$. We therefore have two

Probability and the normal distribution

independent events and the probability of choosing two aces is $(\frac{1}{13} \times \frac{1}{13}) = \frac{1}{169}$.

However, if the initial ace chosen had not been returned to the pack, the probability of choosing a second ace from the fifty one remaining cards must be $\frac{3}{51}$ (or $\frac{1}{17}$) as only three aces are left. This probability of $\frac{1}{17}$ is termed a *conditional probability*. So now the probability of choosing two aces is $(\frac{1}{13} \times \frac{1}{17}) = \frac{1}{221}$.

Example 17.1

There are 200 dwellings in a certain village and they can be classified as follows:

	Small (two bedrooms)	Medium-sized (three bedrooms)	Large (*four or more* bedrooms)
Terraced	32	24	4
Semi-detached	10	68	32
Detached	2	12	16

If an interviewer conducting a survey chooses a dwelling at random, find the probability that it is:

(a) a large or medium-sized detached;
(b) a small or medium-sized dwelling.

If two dwellings are chosen at the same time, find the probability that:

(c) both are large dwellings;
(d) they are both either medium-sized terraced or both large terraced;
(e) they are of different size.

The probability of choosing a particular category of dwelling is given by:

$$\frac{\text{Number in that category}}{\text{Total number of dwellings}}$$

$$\text{eg } P(\text{a large detached}) = \frac{\text{Number of large detached}}{\text{Number of dwellings}}$$

$$= \tfrac{16}{200} = 0.08$$

When one dwelling only is chosen, the events considered are mutually exclusive, so the addition rule applies.

So we can find:

(a) P(a large or medium sized detached)
$= P(\text{a large detached}) + P(\text{a medium sized detached})$
$= \tfrac{16}{200} + \tfrac{12}{200} = 0.14$

(b) P(a small or medium-sized dwelling) = $\frac{44}{200} + \frac{104}{200} = 0.74$
(c) When two dwellings are chosen at the same time, conditional probabilities are involved and the multiplication rule applies.

P(two large dwellings)
= P(a large dwelling) × P(a large dwelling once one has already been chosen)
= $\frac{52}{200} \times \frac{51}{199} = 0.0667$

(d) This involves the addition rule too. We need to find:

P(two medium-sized terraced) + P(two large terraced)
= $(\frac{24}{200} \times \frac{23}{199}) + (\frac{4}{200} \times \frac{3}{199}) = 0.0142$

(e) P(two of different size) = 1 − [P(two small dwellings)
+ P(two medium sized dwellings) + P(two large dwellings)]
= $1 - [(\frac{44}{200} \times \frac{43}{199}) + (\frac{104}{200} \times \frac{103}{199}) + (\frac{52}{200} \times \frac{51}{199})] = 0.6167$

17.3 PERMUTATIONS AND COMBINATIONS

Permutations. A permutation is simply an arrangement of objects. For instance, if there are 9 students in a certain class and there are 4 desks in the front row, how many different arrangements of students in the front seats are possible?

There are 9 students who could fill the first seat, 8 who could have the second, 7 the third and 6 the final one. So the number of ways of filling the four seats is 9 × 8 × 7 × 6 = 3024.

This is the number of permutations of 9 items taken 4 at a time. It is written as P_4^9.

In general we wish to find the number of arrangements of r from n and this is given by

$$P_r^n = \frac{n!}{(n-r)!}$$

where n! (called n factorial) = n × (n − 1) × (n − 2) × $\cdots$ × 1 and
(n − r)! = (n − r) × (n − r − 1) × (n − r − 2) × $\cdots$ × 1

eg $P_4^9 = \dfrac{9 \times 8 \times 7 \times 6 \times \cancel{5} \times \cancel{4} \times \cancel{3} \times \cancel{2} \times \cancel{1}}{\cancel{5} \times \cancel{4} \times \cancel{3} \times \cancel{2} \times \cancel{1}} = 3024$

as already shown.

Combinations. A combination of objects is a selection of n objects taken r at a time without regard to the order in which they are selected.

The formula for a combination is $C_r^n = \dfrac{n!}{r!(n-r)!}$

So if we wished to know the number of ways in which a sub-committee

Probability and the normal distribution

consisting of 2 people could be chosen from a committee of 6 people we could find the answer from

$$C_2^6 = \frac{6!}{2!4!} = \frac{6 \times 5 \times 4 \times 3 \times 2 \times 1}{(2 \times 1) \times (4 \times 3 \times 2 \times 1)} = 15 \text{ ways}$$

17.4 THE BINOMIAL DISTRIBUTION

A binomial distribution, as the name suggests, is concerned with two possible events occurring. For example, if a coin is tossed it will either come down heads or tails.

The formula used for combinations and the multiplication rule can be used together to consider the distribution of probabilities when we have such events.

If p is the probability of obtaining a head when a coin is tossed and q is the probability of obtaining a tail, then when a coin is tossed twice the various possible outcomes are: two heads, one head then one tail, one tail then one head, or two tails. The probabilities of these outcomes being given by p^2, pq, qp and q^2 respectively. As these are the only possible outcomes these probabilities must total 1, ie

$$p^2 + 2pq + q^2 = 1.$$

This is the expansion of $(p + q)^2$

If the coin were tossed n times then the probabilities of the various outcomes would be given by the terms of the expansion of $(p + q)^n$.

This is: $(p + q)^n = p^n + C_1^n \ p^{n-1} \ q + C_2^n \ p^{n-2} q^2 + C_r^n \ p^{n-r} \ q^r + \cdots + C_n^n q^n$ where the term $C_r^n \ p^{n-r} \ q^r$ = the probability of obtaining (n − r) heads and r tails in n tosses of the coin.

For example, if a coin is tossed 6 times the probability of obtaining 4 heads is:

$$\frac{6 \times 5 \times 4 \times 3 \times 2 \times 1}{(4 \times 3 \times 2 \times 1) \times (2 \times 1)} \left(\frac{1}{2}\right)^2 \left(\frac{1}{2}\right)^4 = 0.234$$

This formula can be used even when p and q are not equal. Generally, the mean of a binomial distribution is the number of times (n) an experiment (eg tossing a coin) is carried out times the probability (p) of a success (eg obtaining a head) = np. The standard deviation is $\sqrt{npq}$.

Situations, though, where p and q are both equal are the most interesting. If a coin were tossed 1000 times and the probability of obtaining every possible number of heads from zero to 1000 were calculated, then the frequency curve formed by smoothing a set of histogram blocks depicting the probabilities could be formed as shown in Fig. 17.1.

An important feature of this curve is that it is symmetrical about the mean. Its significance will become apparent in the following sections on the normal distribution.

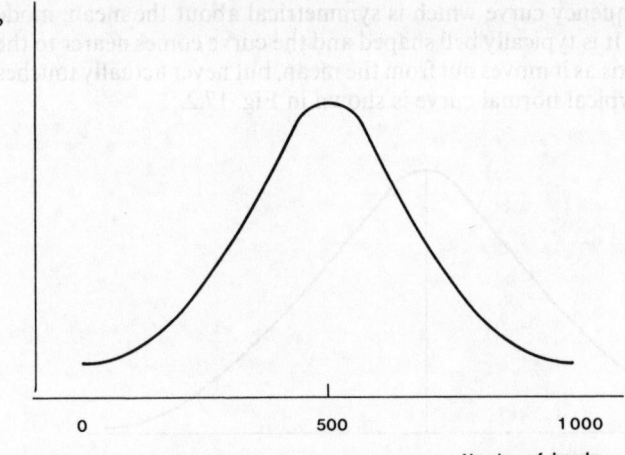

Fig. 17.1 Distribution of probabilities showing the number of heads obtained from 1000 tosses of a coin

17.5 THE NORMAL DISTRIBUTION

As stated in section 16.5.4, it is important when dealing with data to know whether we are dealing with aggregate population data or just a sample, ie a part of the population.

When finding the mean or standard deviation of sample data we are finding a statistic. The mean and standard deviation of the population are called parameters. The distinction is important because the parameters of a population, which includes all possible items, are fixed whereas the statistics obtained from sample data vary according to which items make up the sample.

It is the theoretical distribution of a population with which we are concerned in this chapter.

The notation used to denote the mean and standard deviation of a population differ from that used for sample statistics. Greek letters are used to denote population parameters. The population mean is denoted by μ (pronounced mu) as against the sample mean of $\bar{x}$.

The population standard deviation is σ (pronounced sigma) compared to the sample standard deviation denoted by s.

17.6 THE NORMAL CURVE

The normal distribution (or normal frequency curve) is a theoretical frequency distribution which is of fundamental importance in statistical theory.

Probability and the normal distribution

It is a frequency curve which is symmetrical about the mean, mode and median. It is typically bell shaped and the curve comes nearer to the horizontal axis as it moves out from the mean, but never actually touches the axis. A typical normal curve is shown in Fig. 17.2.

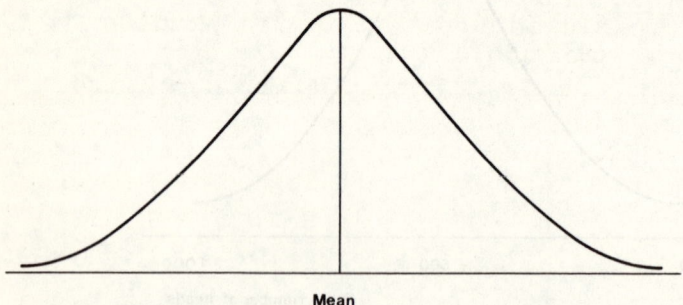

Fig. 17.2 Typical normal curve

The actual shape of any normal curve depends upon the mean and standard deviation of the population which it depicts.

This can be seen from the equation of a normal curve whose mean is μ, and standard deviation is σ. It is

$$y = \frac{1}{\sqrt{2\pi \cdot \sigma^2}} \cdot e^{-\frac{1}{2}((x-u)/\sigma)^2}$$

So if we know the mean and standard deviation of a population we can insert a value for x to find y, ie the height of the curve.

Figure 17.3 shows how two normal distributions can differ in appearance even if they have the same means, but have different standard deviations.

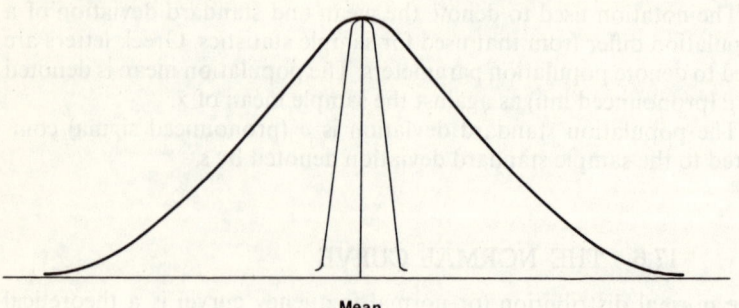

Fig. 17.3 Two normal curves with same means but different standard deviations

To make the normal curve a more useful tool of statistical analysis, it is possible to standardise any normal curve to produce a standard normal distribution.

17.7 THE SIGNIFICANCE OF THE NORMAL DISTRIBUTION

The most important feature of a normal distribution is that the area under the normal curve relates to the proportion of items holding values between two points on the horizontal scale.

For instance, 68.26 per cent of the area of a normal curve lies between the points one standard deviation either side of the mean. Figure 17.4 illustrates the areas found between points one, two and three standard deviations respectively either side of the mean.

If it is known then that an item is normally distributed and the mean and standard deviation values are given, the proportion of the values falling between any two chosen values can be calculated.

To undertake such calculations in practice the task is simplified by the fact that any normal distribution can be converted into a standard normal distribution. The advantage of this is that statistical tables are produced which show the areas under a standard normal curve which fall between certain values.

17.8 THE STANDARD NORMAL DISTRIBUTION

This is a version of the normal distribution which has a mean $(\mu) = 0$ and a standard deviation $(\sigma) = 1$.

The usefulness of the standard normal distribution is that the values in any normal distribution can be converted to values in the standardised version and these converted values looked up in the standard normal distribution table. This conversion is undertaken by changing the scale of the normal distribution.

A value x on the scale of a normal distribution with a mean μ and a standard deviation σ has a corresponding z value on the scale of the standard normal distribution of

$$z = \frac{x - \mu}{\sigma}$$

The resulting z value is called the standard score.

The relationship between the x scale of a normal distribution and the z scale is shown in Fig. 17.5.

So, given any value on the x scale we can convert it to the z scale and once we have converted to the z scale, the table of areas can be used.

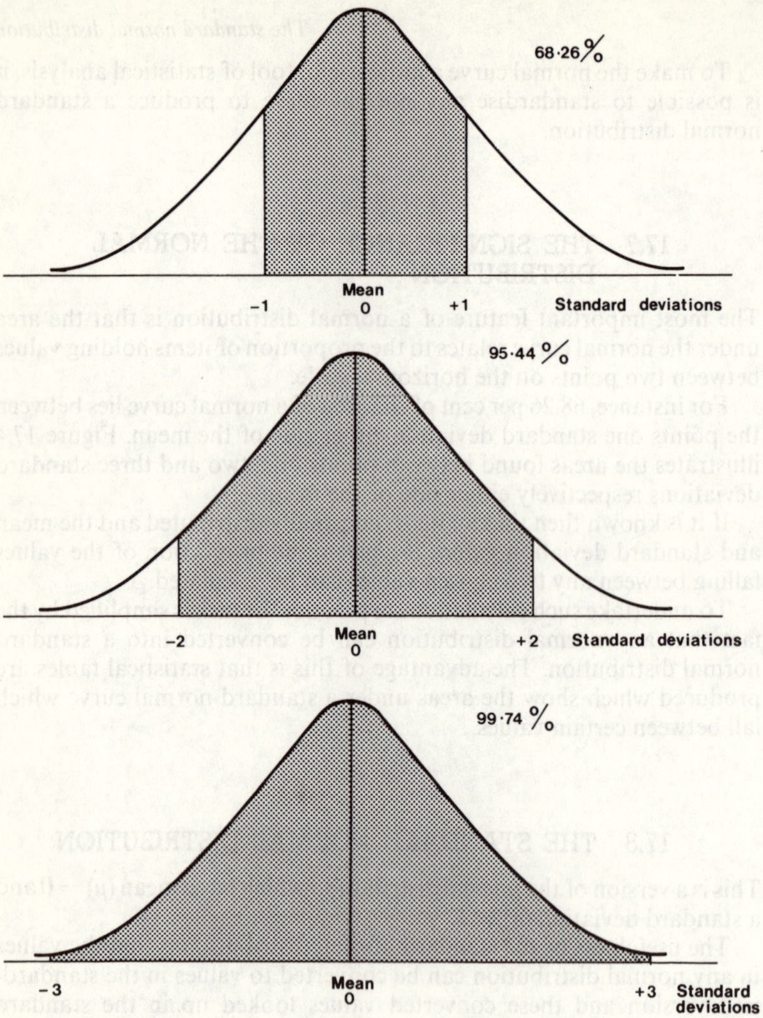

Fig. 17.4 Areas under the normal distribution curve

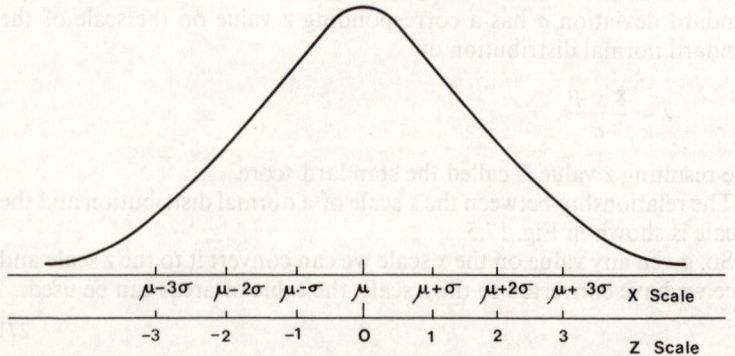

Fig. 17.5 Relationship between the x scale and z scale

17.9 Z VALUES AND AREAS UNDER THE NORMAL CURVE

The area under a normal distribution curve represents all the values of the variable which can arise and the proportion of the area under the curve between any two points on the x scale therefore represents the proportion of the items which hold values between those limits.

A table could be drawn up for any normal distribution with the mean and standard deviation to show such proportionate areas, but as any normal distribution can be standardised, all that is needed is a table which shows the proportionate areas under the standard normal curve. Such a table is shown in the Appendix (p. 346). The total area under the standard normal curve is equal to 1 (or 100 per cent) as it contains all the items in the population and the table is designed to show the proportion of the area under the curve which is to be found to the left of any z value.

(It is an unfortunate fact that some statistical tables of the normal distribution are composed in a different way and might instead give the area under the curve which is to the right of the z value. The user of any normal curve table should check carefully exactly what is being measured in any table being used.)

The way to use the 'z table' in the Appendix is to look down the z column to find the calculated z value and then read off the area which is given as a decimal figure in the next column.

Examples on finding areas.

1. As the mean of the standard normal distribution is 0 (see Fig. 17.5) then obviously exactly 50 per cent of the area under the curve lies to the left of the mean, and so the area which corresponds to a z value of 0.00 is given as 0.5000 (ie 50 per cent).

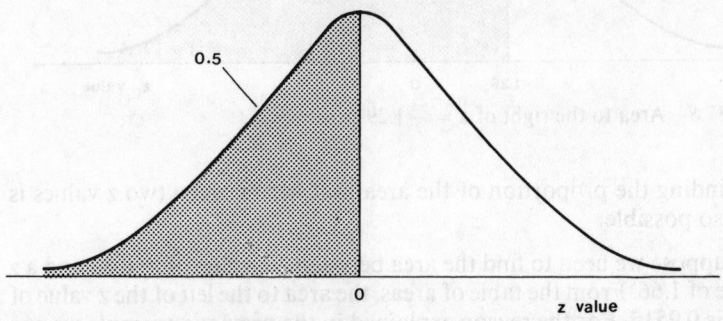

Fig. 17.6 Area to the left of z = 0.00

2. For a positive z value it is a simple matter to use the z table. The area which lies to the left of the z value of 1.29 is 0.9015, which means that just over 90 per cent of the area lies to the left of 1.29.

Probability and the normal distribution

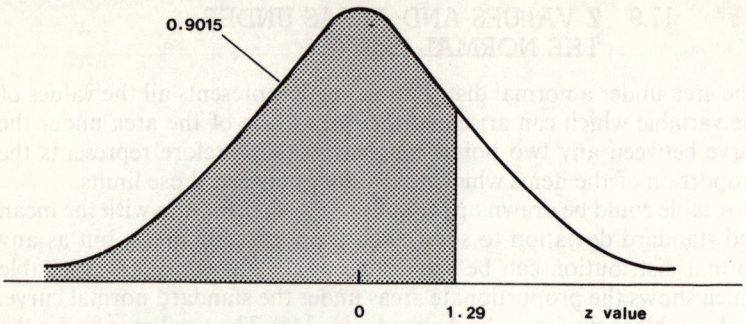

Fig. 17.7 Area to the left of z = 1.29

3. If the calculated value of z is negative, the corresponding positive value can be looked up in the table, and due to the exact symmetry of the normal curve, the area to the left of this positive z value corresponds to the area to the right of the negative z value, ie if the calculated z value is −1.29 the proportion of the area to the *right* of this value is just over 90 per cent of the total area.

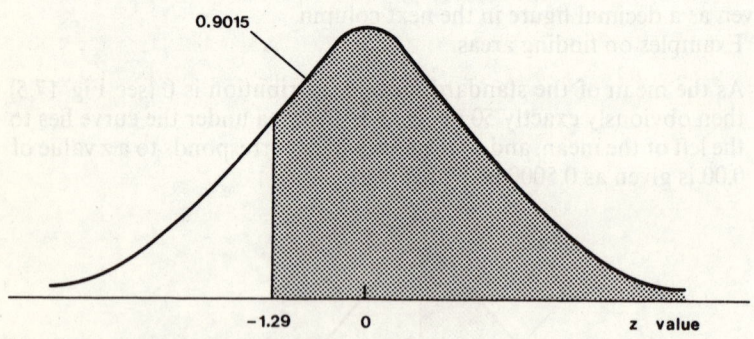

Fig. 17.8 Area to the right of z = −1.29

4. Finding the proportion of the area that lies between two z values is also possible.

Suppose we need to find the area between a z value of −0.82 and a z value of 1.66. From the table of areas, the area to the left of the z value of 1.66 is 0.9515. For the reason explained in the previous example we can find from the table the area to the right of z = −0.82. It is 0.7939. So the area to the *left* of the z value of −0.82 must be 1 − 0.7939 = 0.2061.

Subtracting 0.2061 from 0.7939 gives 0.5878 as the shaded area in Fig. 17.9, ie 58.78 per cent of the area lies between the z values of −0.82 and 1.66.

Areas under the normal curve and probabilities

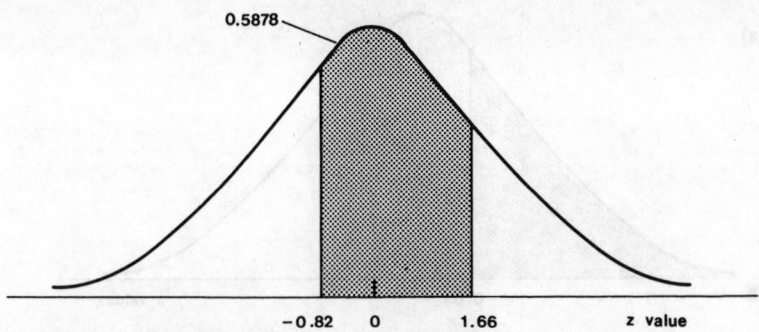

Fig. 17.9 Area between z = −0.82 and z = 1.66

17.10 AREAS UNDER THE NORMAL CURVE AND PROBABILITIES

To understand the use of the normal curve in the calculation of probabilities, we can relate back to section 17.4 on the binomial distribution.

Any area under the binomial curve between two points depicts the probability of obtaining values between those points and the same is true for the normal curve.

If we know that a variable is normally distributed, then the area under the curve between any two points indicates the probability of a randomly chosen item holding a value between those two points.

The following example shows this.

Example 17.2

The heights of a town's adult male population are known to be normally distributed with a mean of 1.75 m and a standard deviation of 0.12 m.

If an individual is chosen at random from the population find the probability that he will be:

(a) under 1.65 m;
(b) over 1.95 m;
(c) between 1.70 m and 1.85 m.

(a) $z = \dfrac{1.65 - 1.75}{0.12} = \dfrac{-0.10}{0.12} = -0.83$

We need to find the area to the left of z = −0.83 as shown in Fig. 17.10(a).

From the z table, the area to the left of z = +0.83 is 0.7967. So the area to the right of z = −0.83 is also 0.7967. This means that the required area is (1 − 0.7967) = 0.2033.

The probability of a randomly chosen individual being under 1.65 m is therefore 0.2033.

Probability and the normal distribution

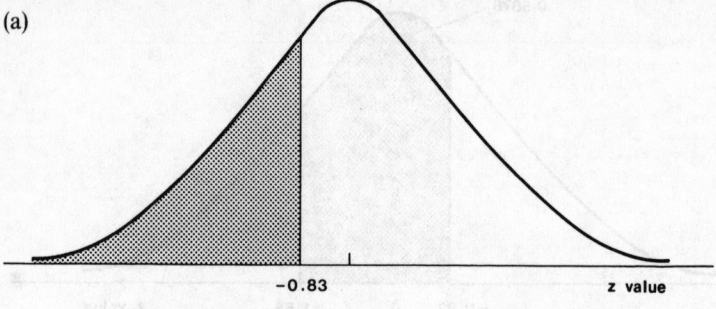

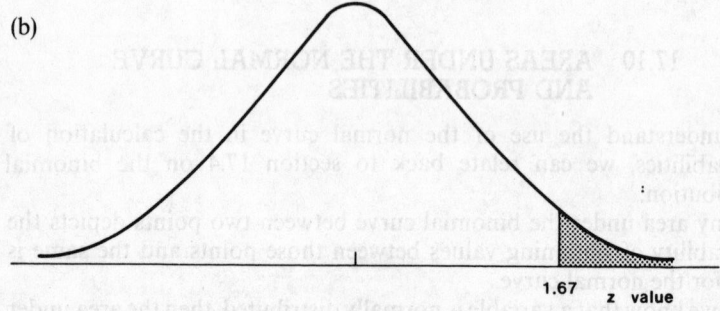

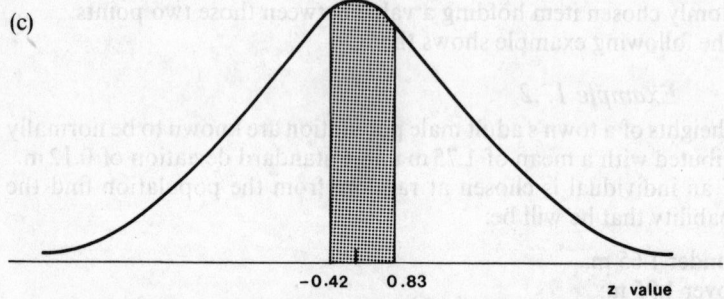

Fig. 17.10 (a) Area to the left of z = −0.83; (b) area to the right of z = 1.67; (c) area between z = −0.42 and z = 0.83

(b) $z = \dfrac{1.95 - 1.75}{0.12} = \dfrac{0.20}{0.12} = 1.67$

The area to the right of 1.67 (see Fig. 17.10b) is required.

From the z table, the area to the left of z = 1.67 is 0.9525. The required area is thus (1 − 0.9525) = 0.0475. This is the probability of a chosen individual being taller than 1.95 m.

276

(c) Here we need to find two z values. z_1 is the value which corresponds to 1.70 m and z_2 relates to 1.85 m.

$$z_1 = \frac{1.70 - 1.75}{0.12} = \frac{0.05}{0.12} = -0.42$$

and $$z_2 = \frac{1.85 - 1.75}{1.12} = \frac{0.10}{0.12} = 0.83$$

The area we need to find is that between z_1 and z_2 (see Fig. 17.10c). From the z table the area to the left of $z = +0.42$ is found to be 0.6628 and so the area to the right of $z = -0.42$ is also 0.6628. This means that the area to the left of $z = -0.42$ must therefore be $(1 - 0.6628) = 0.3372$.

Now the area to the left of $z = 0.83$ is given as 0.7967 in the table. The required area, ie that between $z_1 = -0.42$ and $z_2 = 0.83$, must be equal to $(0.7967 - 0.3372) = 0.4595$, which is the probability that the chosen individual is between 1.70 m and 1.85 m.

17.11 EXERCISES

1. A library contains 75 000 books of which 1500 are relevant to quantity surveying. It also contains 1000 journals of which twenty-five are relevant to quantity surveying. A man enters the library, selects a book at random and then looks at a journal at random. What is the probability that:

   (a) both book and journal are relevant to quantity surveying;
   (b) one is relevant but the other is not?
   While he is reading the journal a second man enters and picks a journal at random, what is the probability that
   (c) both are quantity surveying journals;
   (d) neither is a quantity surveying journal? (RICS)

2. From observations over a long period of time the following probabilities have been found for three machines on a building site.

	Working	Idle	Broken down
Machine 1	0.6	0.3	0.1
Machine 2	0.7	?	0.2
Machine 3	?	0.4	0.1

Use the laws of probability to find:

(a) the missing values;
(b) the probability that all machines are working;

(c) the probability that no machines are broken down;
(d) the probability that at least two machines are idle. (RICS)
3. Direct access to a main-frame computer can be made from one of three rooms, P, Q and R. Each room has two types of terminal, teletypes and visual display units. The numbers are given below.

Room	P	Q	R
Teletypes	15	12	10
VDU	5	8	10

(a) If the probabilities that a user goes to room P, Q or R are 0.6, 0.3 and 0.1 respectively and then selects a terminal at random, what is the probability that it is a teletype?
(b) If two users go to a room at random and also select a terminal at random, what is the probability that they both use the same type of machine? (RICS)

18 INDEX NUMBERS

18.1 INTRODUCTION

An index number is a measure, over time, designed to show average changes in the price, quantity or value of a group of items.

If we wish to compare several series of figures it is more than likely that their complexity will render direct comparison meaningless. If, for instance, we had information on every form of production in the building industry during this year and last year (eg number of bricks produced, tonnes of cement, etc), the sheer mass of data would make it impossible to see clearly in which year production was higher. Instead of such an excess of figures, what we need is a single figure, which in itself shows how much one year differs from another. A convenient way of doing this is to take a typical year's figures as a base and express the figures for other years as a measure of this. Such a single figure, summarising a comparison between the two sets of figures, is called an index number.

Although it has just been stated that index numbers are mainly used to deal with a number of items, the simplest type of index number series to compile is a single item index.

18.2 SINGLE-ITEM INDEX NUMBERS

Where only one item is involved in comparisons between different periods, the calculation of index numbers is very simple.

Assuming that annual data are being dealt with, one year is chosen as base and the values for the other years are stated in proportion to the value for the base year.

Index numbers

Example 18.1

A firm of builder's merchants has the following sales figures for a particular commodity as shown in Table 18.1.

Year	Amount sold (*tonnes*)
1974	430
1975	472
1976	483
1977	493
1978	502
1979	510

If 1975 is chosen as the base year, then the index number for 1975 is 100, and the index number for any other year is found by dividing that year's sales figure by the 1975 sales figure and then multiplying by 100 to put the resulting proportion into index number form. This can be expressed by the notation;

$$\text{Index number for year n} = \frac{q_n}{q_0} \times 100$$

Where

q_0 = the quantity sold in the base year (1975 here)

q_n = the quantity sold in the current year.

The whole index number series can thus be calculated as in Table 18.2.

Table 18.2 Calculation for whole index number series

Year	Amount sold (*tonnes*)	Index of sales
1974	430	430/472 × 100 = 91.1
1975	472	100
1976	483	483/472 × 100 = 102.3
1977	493	493/472 × 100 = 104.4
1978	502	502/472 × 100 = 106.6
1979	510	510/472 × 100 = 108.0

Table 18.3 Quantities and prices of commodities sold by builder's merchants

Commodity	Unit	(1976) Quantity	Price per unit (£)	(1977) Quantity	Price per unit (£)	(1978) Quantity	Price per unit (£)
A	tonne	120	10.0	150	10.4	160	11.0
B	tonne	50	20.0	60	20.0	70	24.0
C	m^2	1000	4.0	1200	4.2	1200	4.8
D	m^3	500	4.0	700	4.6	800	5.0

Index numbers

18.3 MULTI-ITEM INDEX NUMBERS

Where more than one item is involved a multi-item index needs to be used. This normally means the calculation of a weighted aggregative index.

Example 18.2

A firm of builder's merchants divides its commodities into four main groups and the quantities sold and its prices (£), for the past three years are as shown in Table 18.3.

The firm wishes to measure *the overall* change in prices over this period and 1976 is chosen as the base year.

In other words, the firm wishes to find one index figure for each year which shows how the prices of its products have changed in aggregate: it does not wish to produce an index for each commodity individually.

In constructing its index then the firm must take into account the relative importance of the various commodities, eg the fact that much more of commodity 'C' than of commodity 'D' is sold each year, so the effect of a change in price of the former commodity will be so much greater. A weighted average must therefore be found.

The obvious weightings to take in this example are the quantities of the commodities sold. One way of doing this is to use weights which show the relative importance of the commodities in the base period, ie the base year quantities. Such an index is said to be 'base weighted' or of the **Laspeyres** type.

Another method is to calculate the weights using quantities consumed in the current year, and an index number of this type is described as 'current weighted' or a **Paasche** type.

These two types of index are defined as follows:

$$\text{Laspeyres price index for year n} = \frac{\Sigma p_n q_0}{\Sigma p_0 q_0} \times 100$$

Where

p_n and p_0 are the current and base year prices respectively, q_n and q_0 are the respective quantities.

This index indicates how much the cost of buying base year quantities at current year prices compares with base year costs.

$$\text{Paasche price index for year n} = \frac{\Sigma p_n q_n}{\Sigma p_0 q_n} \times 100$$

This index indicates how much current year costs are related to the cost of buying current year quantities at base year prices.

Returning to our example, we can now calculate these indices.

The calculations for Laspeyres price index with 1976 (= 100) as base year are as follows:

Comparison of Laspeyres and Paasche indices

$$\Sigma p_{1976} q_{1976} = (10.0 \times 120) + (20.0 \times 50)$$
$$+ (4.0 \times 1000) + (4.0 \times 500)$$
$$= 8200$$
$$\Sigma p_{1977} q_{1976} = (10.4 \times 120) + (20.0 \times 50)$$
$$+ (4.2 \times 1000) + (4.6 \times 500)$$
$$= 8748.$$
$$\Sigma p_{1978} q_{1976} = (11.0 \times 120) + (24.0 \times 50)$$
$$+ (4.8 \times 1000) + (5.0 \times 500)$$
$$= 9820$$

The price index series for the three years is:

1976   = 100
1977 8748/8200 × 100 = 106.7
1978 9820/8200 × 100 = 119.8

Paasche price index with 1976 (= 100) as base year

$$\Sigma p_{1976} q_{1977} = (10.0 \times 150) + (20.0 \times 60)$$
$$+ (4.0 \times 1200) + (4.0 \times 700)$$
$$= 10\,300.$$
$$\Sigma p_{1977} q_{1977} = (10.4 \times 150) + (20.0 \times 60)$$
$$+ (4.2 \times 1200) + (4.6 \times 700)$$
$$= 11\,020.$$
$$\Sigma p_{1976} q_{1978} = (10.0 \times 160) + (20.0 \times 70)$$
$$+ (4.0 \times 1200) + (4.0 \times 800)$$
$$= 11\,000.$$
$$\Sigma p_{1978} q_{1978} = (11.0 \times 160) + (24.0 \times 70)$$
$$+ (4.8 \times 1200) + (5.0 \times 800)$$
$$= 13\,200.$$

The price index series for the three years is:

1976   = 100
1977 11 020/10 300 × 100 = 107.0
1978 13 200/11 000 × 100 = 120.0

18.4 COMPARISON OF LASPEYRES AND PAASCHE INDICES

The results which we obtain by each method would only differ substantially if there is any great change in the pattern of sales.

One major advantage which the Laspeyres price index has is in the ease of computation; the denominator of the formula remains the same

Index numbers

each period ($\Sigma p_0 q_0$), whereas with the Paasche price index ($\Sigma p_0 q_n$) it has to be recalculated every period. Also, as a result of this, the different years in the Laspeyres index can be directly compared with each other and not just the base year.

The Paasche price index, on the other hand, is much more useful when the quantities of each commodity being sold each year are likely to vary to any degree. It may well be the case, that due to some technological advance, there is an increase in demand for a certain commodity each year, and a base quantity weighted price index would not pick this up.

18.5 PRICE RELATIVES INDEX

A price relative is simply the price of an item in one year relative to another year, again using 100 as base.

It can be depicted, using our symbols as:

$$\frac{p_n}{p_0} \times 100$$

So, in our previous example, commodity A has a price relative of $\frac{10.4}{10.0} \times 100 = 104$, comparing 1977 with 1976.

If we are concerned with the compilation of a multi-item index, a composite index number can be obtained by taking a weighted average of all the price relatives with which we are concerned.

The formula for this index number is:

$$\frac{\Sigma \left(\frac{p_n}{p_0} \times \text{weight} \right)}{\Sigma \text{weights}} \times 100$$

Table 18.4 Price relatives index

Commodity	1976 Price per unit (£)	1977 Price per unit (£)	Price relative	Weight	Price relative × weight
A	10.0	10.4	104	6	624
B	20.0	20.0	100	5	500
C	4.0	4.2	105	20	2100
D	4.0	4.6	115	10	1150
				41	4374

$$\text{Index for 1977} = \frac{4374}{41} = 106.7$$

This is as was previously calculated in section 18.3.

284

Choice of base year

We can, for instance, find a price relatives index for 1977 for Example 18.2 using the values of sales for 1976 as relative weights. This is shown in Table 18.4.

It should be noted that the value of sales in the base year ($p_0 q_0$) should be used to formulate the weightings rather than the base year quantities sold. (The reader may wish to work out the reason for this, using our algebraic notation.) The base year values of sales are:

A: £1200   B: £1000   C: £4000   D: £2000.

These can be reduced to 6, 5, 20 and 10 respectively.

It is to be expected that this index gives the same result as previously.

The main use of separate calculations of price relatives is that they allow a comparison of the relative change compared with the base data for each item.

18.6 CHOICE OF BASE YEAR

With a fixed base index, it is clearly important that the base year is selected so as to provide a satisfactory standard of comparison. A high base year value tends to detract from other years and a low one tends to make them appear higher. For this reason an average of several years is sometimes used to form an artificial base.

As time moves on the base may become unrealistic. This can occur if there has been a pronounced inflationary trend in the data so that, say, the base year prices bear little relation to current prices. This may be of little importance if we are prepared to put up with a very large index number, but in practice a conversion to a new base year may well occur as happened with the UK Retail Price Index in 1974.

Conversion may be made to a new year as base simply by dividing all the index numbers by the index of the new base year on the old scale and multiplying by 100.

In Example 18.1 we could change the base of the index to 1977 = 100 as shown in Table 18.5.

Table 18.5 Change of base year

Year	Present index of sales (1975 = 100)	New index of sales (1977 = 100)
1974	91.1	91.1/104.4 × 100 = 87.3
1975	100	100/104.4 × 100 = 95.8
1976	102.3	102.3/104.4 × 100 = 98.0
1977	104.4	100
1978	106.4	106.4/104.4 × 100 = 101.9
1979	108.0	108/104.4 × 100 = 103.9

Index numbers

A more serious objection to index number comparisons over long periods of time is that the basic circumstances change. Even if we are dealing with what seems to be a single commodity, it rarely remains exactly the same. Designs change and quality varies. This applies especially if the index number relates to a number of items, such as, for example, in the Retail Price Index.

18.7 CHAIN-BASE INDEX NUMBERS

A chain-base index is simply one in which each period in the series uses the previous period as a base.

Taking Example 18.1, a chain-base index with 1974 as the initial period would be calculated as shown in Table 18.6.

Table 18.6 Calculations for chain-base index

Year	Amount sold (tonnes)	Chain base index
1974	430	100
1975	472	472/430 × 100 = 109.8
1976	483	483/472 × 100 = 102.3
1977	493	493/483 × 100 = 102.0
1978	502	502/493 × 100 = 101.8
1979	510	510/502 × 100 = 101.6

Such an index shows whether the rate of change is rising, falling or constant, as well as the extent of the change from year to year. It can be seen in this example that although there is a steady increase in the rise in sales, the increase each year in relation to the total sales of previous year is falling.

The main advantage of a chain-base index is the fact that new items can be easily accommodated into the index.

18.8 USE OF INDEX NUMBERS

There are several important considerations to bear in mind when index numbers are being studied. The purpose of the index is of overall importance. The Retail Price Index, for example, may be considered to be a measure of the cost of living in this country, but officially, as the title of the index suggests, it merely measures changes in certain retail prices. If it were to be considered truly to be a cost of living index, then certain other items such as personal income tax may need to be included in the index.

The choice of weights for an index will obviously determine the resulting index number for each year. As we have seen from a comparison of index numbers by the Laspeyres and Paasche methods, if the sales figures in Example 18.2 were to vary considerably from year to year, the resulting index number series under each method may differ a great deal.

Lastly the base year of the index, especially for a price index in inflationary periods, will hopefully not be so far distant that it is difficult to relate the figures back. If this is a problem, then a rebasing might be needed.

18.9 EXERCISES

1. (a) Why is weighting used in the construction of index numbers? Describe how weighting is achieved with Laspeyres and Paasche prices indices.
   (b) Construct a weighted index series from the following data to show (i) the change in prices and (ii) the change in quantities

Item	Unit	Quantity			Price (£)		
		1977	1978	1979	1977	1978	1979
X	m^2	1000	1000	1200	6	7	9
Y	m^3	400	500	600	10	12	15
Z	tonne	100	150	200	10	10	14

2. What are the main factors to be considered in constructing an index number series?
   A firm classifies its output into three types of building A, B, and C. Its records for the past four years are:

	1969		1970		1971		1972	
Type	a	b	a	b	a	b	a	b
A	6	120	8	168	9	198	11	264
B	10	100	12	132	16	176	20	240
C	40	240	44	286	50	315	60	387

a is the number of units.   b is the value in £000s.

Index numbers

Use the data to compile an index number series with 1969 as base year, using a weighted arithmetic mean. The purpose of the index is to measure the overall change in quantity of buildings produced.

(RICS)

3. (a) Explain the purpose of weighting in index number construction.
   (b) Summarise the advantages and disadvantages of base-weighted and of current-weighted index numbers.
   (c) Combine the two following index number series into a base-weighted (Laspeyres), average-of-relative type index series, based on 1978.

Mid-year	Labour costs		Material costs	
	Index	Weight	Index	Weight
1978	196	41	198	59
1979	238	48	236	52
1980	282	47	283	53

(RICS)

288

19 LINEAR REGRESSION

19.1 INTRODUCTION

In business or industry it is often necessary to predict trends and it is the job of the statistician to make these predictions.

Predictions based on statistical information can only be in terms of probability but they are satisfactory if a high percentage of them prove to be correct.

We usually need to express relationships between quantities that are known and quantities that are to be predicted by using mathematical equations.

The simplest and most widely used equation for expressing such relationships is the linear equation

$$Y = a + bX$$

where a and b are constants, and if we know the values of a and b then we can predict the corresponding value for the variable Y for a given value of variable X.

19.2 A LINEAR RELATIONSHIP

If the producer of a certain product X has fixed costs of production of £10 000 and constant variable costs of £3000 per unit of output of X then the total cost (Y) for any level of output can be found from the equation,

$$Y = £(10\,000 + 3000X)$$

Linear regression

So if an output of, say, 5 units of X is produced the total costs will be

$$Y = £10\,000 + (3000 \times 5)$$
$$= £25\,000$$

This linear relationship is depicted in Fig. 19.1.

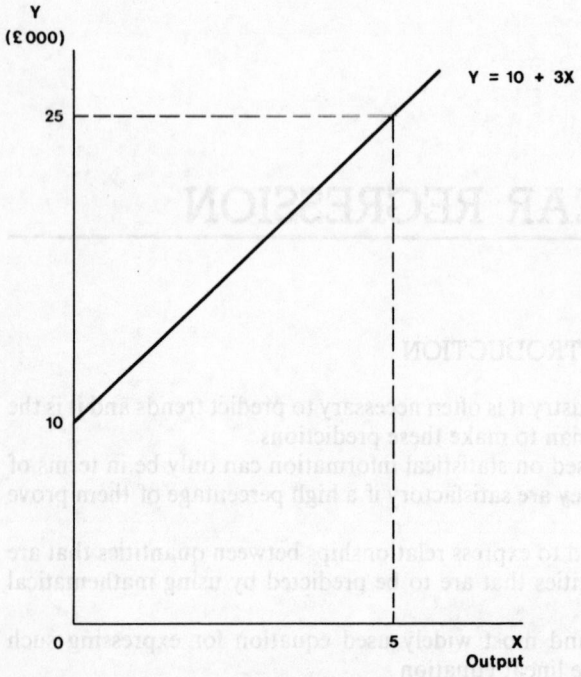

Fig. 19.1 Relationship between costs and output

In many situations, though, we may have sets of values for two variables and we may believe that there is a linear relationship between the two lots of data but we do not know the value of the two parameters a and b. If we are not given the values of a and b we need some method of calculating them so as to fit appropriately the data we are given.

Example 19.1

A supplier of building materials has information on his advertising expenditure on a certain product and also the returns from the sale of that product over the previous eight years. Using this information (shown in Table 19.1) the supplier would like to estimate his sales figure for a future year after he has decided on his advertising expenditure.

A linear relationship

The two sets of values shown in Table 19.1 are plotted on a diagram called a **scatter diagram,** as shown in Fig. 19.2.

Table 19.1 Advertising expenditure and sales returns

Year	Advertising (£00)	Sales (£0000)
1	1	1
2	3	2
3	4	4
4	6	4
5	8	5
6	9	7
7	11	8
8	14	9

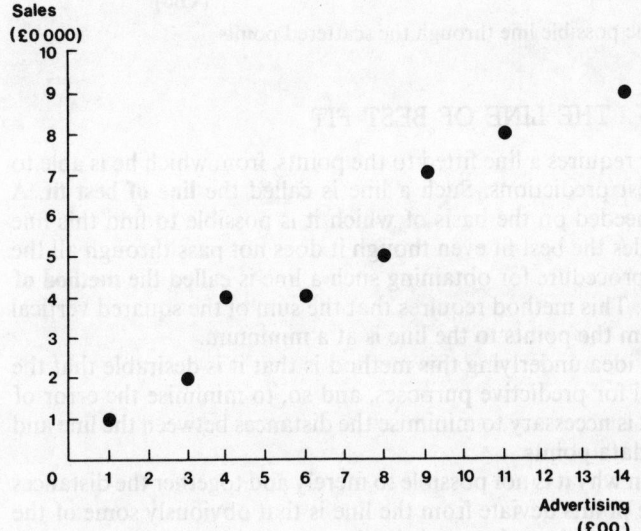

Fig. 19.2 Scatter diagram of sales and advertising data

It would appear from the diagram that the relationship between these two variables may approximate to a linear one. A line can be drawn through the points as in Fig. 19.3, but it is only one of many that can be drawn and the problem is one of choosing which is the best line to fit to the points.

Linear regression

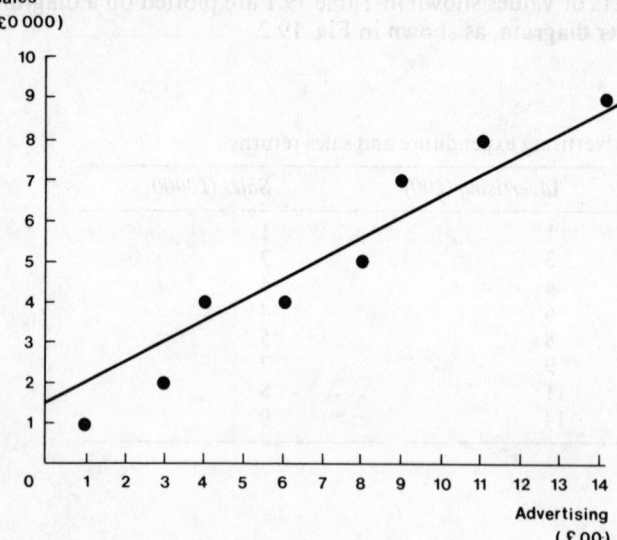

Fig. 19.3 One possible line through the scattered points

19.3 THE LINE OF BEST FIT

The supplier requires a line fitted to the points, from which he is able to make the best predictions. Such a line is called the **line of best fit**. A criterion is needed on the basis of which it is possible to find this line which provides the best fit even though it does not pass through all the points. The procedure for obtaining such a line is called the **method of least squares**. This method requires that the sum of the squared vertical distances from the points to the line is at a minimum.

The basic idea underlying this method is that it is desirable that the line be useful for predictive purposes, and so, to minimise the error of prediction, it is necessary to minimise the distances between the line and the existing data points.

The reason why it is not possible to merely add together the distances by which the points deviate from the line is that obviously some of the distances between the points and the line are positive whilst others are negative. Therefore the sum of these distances could be quite small even when the deviations are large. Such problems are overcome by squaring the deviations.

19.4 THE METHOD OF LEAST SQUARES

The standard approach to this method is to work from what are called normal equations. These are two simultaneous equations which when

The method of least squares

solved using the sample data come up with values for the two constants a and b.

These equations are:

$$\Sigma Y = an + b\Sigma X$$

and $\quad \Sigma XY = a\Sigma X + b\Sigma X^2$

where n is the number of pairs of observations and ΣX, ΣY, ΣXY and ΣX^2 are calculated from the data.

These values can be substituted in the two equations which can then be solved simultaneously to give the required values of a and b.

By solving the above normal equations the equation for the regression line of Y on X is found. This means that the line found can suitably be used to enable the prediction of a value for Y from a known value of X. Therefore, X is said to be the independent variable while Y is the dependent variable. When labelling the variables it is of vital importance that this distinction is borne in mind.

A regression line of X on Y could also be found by interchanging the Xs and Ys in the normal equations and performing a separate calculation, but the two regression lines found would not be the same and one could not be obtained directly from the other.

In our example, then, the advertising figures will be the X values and the sales figures the Y values, as the supplier is trying to find an equation which will enable him to predict the level of sales, given a certain advertising expenditure.

The calculations as shown in Table 19.2 must therefore be carried out.

We have the following values:

$$\Sigma X = 56 \quad \Sigma Y = 40 \quad \Sigma X^2 = 524 \quad \Sigma XY = 364 \text{ and } n = 8.$$

Putting these values into the normal equations:

$40 = 8a + 56b \qquad\qquad 1$

$364 = 56a + 524b \qquad\qquad 2$

Table 19.2 Calculations for finding regression equation

Year	Advertising (£00) X	Sales (£0000) Y	X^2	XY
1	1	1	1	1
2	3	2	9	6
3	4	4	16	16
4	6	4	36	24
5	8	5	64	40
6	9	7	81	63
7	11	8	121	88
8	14	9	256	126
	56	40	524	364

Linear regression

In order to eliminate the a's multiply equation 1 by 7.

$$280 = 56a + 392b \qquad 3$$

Subtracting equation 3 from equation 2,

$$84 = 132b$$

$$\therefore b = \frac{84}{132} = 0.636$$

Substituting this value for b into equation 1 gives

$$40 = 8a + (56 \times 0.636)$$

$$\therefore 8a = 40 - 35.6 = 4.4$$

$$\therefore a = \frac{4.4}{8} = 0.55$$

The regression equation is

$$Y = 0.55 + 0.636X$$

The regression line is shown in Fig. 19.4. It can be drawn quite simply from the equation by inserting two values for X into the equation, finding the corresponding value for Y in each case, plotting the resultant points and joining the points by a straight line.

19.5 USE OF THE REGRESSION LINE FOR PREDICTION PURPOSES

Any value for X can now be substituted into the regression equation in order to give the estimated level of Y. If the X value used is within the range of values for which data already exist then to estimate the corresponding value for Y is termed **interpolation**.

If, however, the regression line is extended and some value for X is chosen which has never been met with before, then the estimated value for Y is found by **extrapolation**.

For instance, if our supplier wished to estimate the level of sales he could expect from an expenditure of £700 on advertising, this level of expenditure lies within the existing range of advertising figures (£100–£1400) and so the corresponding sales figure can be found by interpolation.

To find the likely level of sales associated with a higher level of advertising expenditure than he has undertaken before, eg £1600, extrapolation is needed.

These estimated sales figures can be found quite simply by plugging the advertising values into the regression equation.

When £700 is spent,

$$X = 7 \text{ and}$$
$$Y = 0.55 + (0.636 \times 7)$$
$$= 5.002$$

ie an estimated sales value of £50 020.

When £1600 is spent

$$X = 16 \text{ and}$$
$$Y = 0.55 + (0.636 \times 16)$$
$$= 10.726$$

ie an estimated sales value of £107 260.
These figures could also be read off from the regression line.

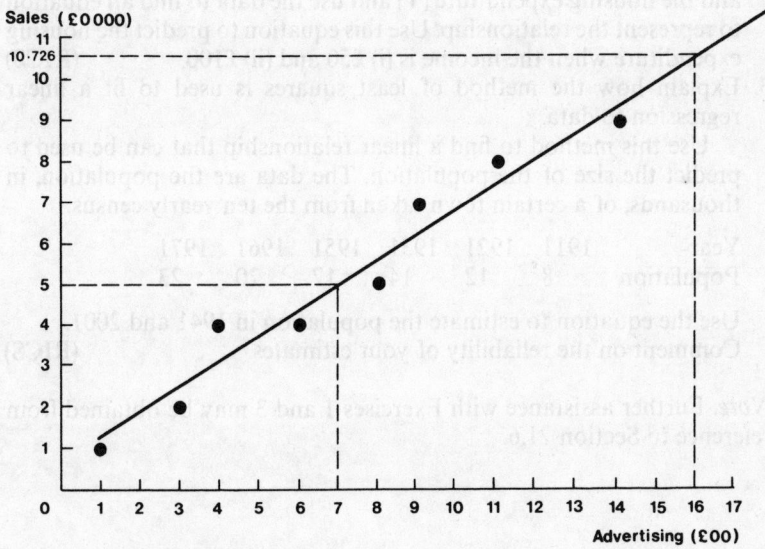

Fig. 19.4 Regression line

19.6 EXERCISES

1. (a) By reference to the data in the table demonstrate how an estimate of the linear trend in construction costs could be derived.
   (b) Suggest what further analysis might be desirable in view of the results obtained.

Linear regression

Year	Out turn tender price indices (*Private commercial*)
1970	100
1971	110
1972	126
1973	158
1974	220
1975	266
1976	285
1977	305

(RICS)

2. A household expenditure survey produced the following data.

Income (£)	45	55	65	75	85
Housing expenditure (£)	7.22	7.47	8.53	8.68	9.71

Assume that there is a linear relationship between the income (X) and the housing expenditure (Y) and use the data to find an equation to represent the relationship. Use this equation to predict the housing expenditure when the income is (i) £50 and (ii) £100. (RICS)

3. Explain how the method of least squares is used to fit a linear regression to data.

Use this method to find a linear relationship that can be used to predict the size of the population. The data are the population, in thousands, of a certain town taken from the ten yearly census.

Year	1911	1921	1931	1951	1961	1971
Population	8	12	14	17	20	23

Use the equation to estimate the population in 1941 and 2001. Comment on the reliability of your estimates. (RICS)

Note. Further assistance with Exercises 1 and 3 may be obtained from reference to Section 21.6.

20 CORRELATION

20.1 INTRODUCTION

So far we have fitted a least squares regression line to our sample observations. From the examples we have done, the positions of the points on the graph have made it seem reasonable to fit a straight line to the points. In some cases, however, we might be doubtful as to whether predictions using a straight line fitted to the points will be a true representation and as to whether predictions using a straight line will be accurate.

It is therefore necessary to have some measure of the accuracy of fitting a straight line through our points.

20.2 SCATTER DIAGRAMS

If we take a scatter diagram (Fig. 20.1),

$$\text{Let } x_i = X_i - \bar{X}$$
$$y_i = Y_i - \bar{Y}$$

In quadrant I,

$$(X_i - \bar{X}) > 0 \quad \text{and} \quad (Y_i - \bar{Y}) > 0$$
$$\therefore \quad x_i y_i > 0$$

In quadrant II,

$$(X_i - \bar{X}) < 0 \quad \text{and} \quad (Y_i - \bar{Y}) > 0$$
$$\therefore \quad x_i y_i < 0$$

Correlation

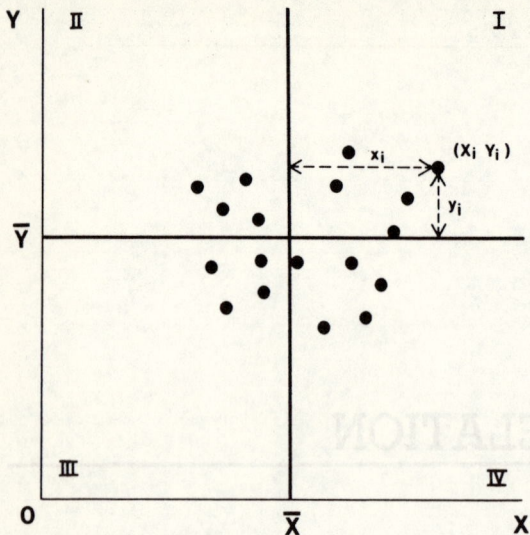

Fig. 20.1 Scatter diagram

In quadrant III,

$(X_i - \bar{X}) < 0$ and $(Y_i - \bar{Y}) < 0$
$\therefore\ x_i y_i > 0$

In quadrant IV,

$(X_i - \bar{X}) > 0$ and $(Y_i - \bar{Y}) < 0$
$\therefore\ x_i y_i < 0$

Consider the three scatter diagrams shown in Fig. 20.2.

In Fig. 20.2(a), as the association between X and Y is positive most of the points lie in quadrants I and III and $\Sigma x_i y_i$ will tend to be positive.

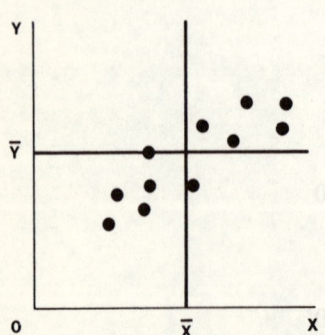

Fig. 20.2 (a) Positive relationship between X and Y;

In Fig. 20.2(b), as the association is negative, most of the points will lie in quadrants II and IV and $\Sigma x_i y_i$ will tend to be negative.

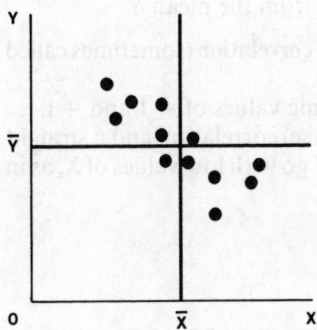

Fig. 20.2 (b) Negative relationship between X and Y;

In Fig. 20.2(c), if there is no obvious association between X and Y, the points will be scattered over all four quadrants and $\Sigma x_i y_i$ will tend to be small as the negative values largely balance out with the positive values.

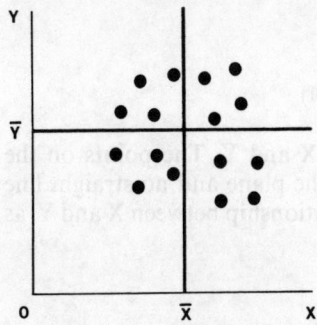

Fig. 20.2 (c) No relationship between X and Y

20.3 THE PRODUCT MOMENT CORRELATION COEFFICIENT

The value $\Sigma x_i y_i$ is affected by the scale of measurement of X and Y and also by the number of observations. An ideal measure is not affected by these factors.

Such a measure is

$$r = \frac{\Sigma x_i y_i}{\sqrt{(\Sigma x_i^2 \Sigma y_i^2)}}$$

Correlation

where

x_i = the deviation of the X_i value from the mean $\bar{X}$

y_i = the deviation of the Y_i value from the mean $\bar{Y}$

r is called the **product moment coefficient of correlation** (sometimes called Pearson's coefficient of correlation).

The value of r varies between the extreme values of -1 and $+1$.

If **r = −1** there is perfect negative (or inverse) correlation and a straight line exactly fits the points. High values of Y go with low values of X, as in Fig. 20.3.

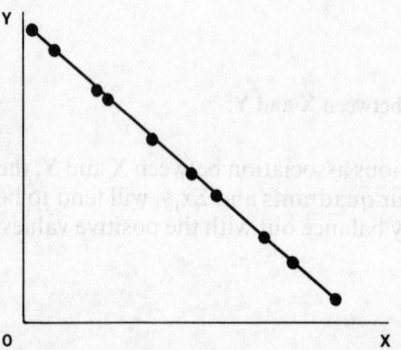

Fig. 20.3 Perfect negative correlation (r = −1)

If **r = 0** there is no correlation between X and Y. The points on the diagram would be scattered throughout the plane and no straight line could be said to properly represent the relationship between X and Y, as in Fig. 20.4.

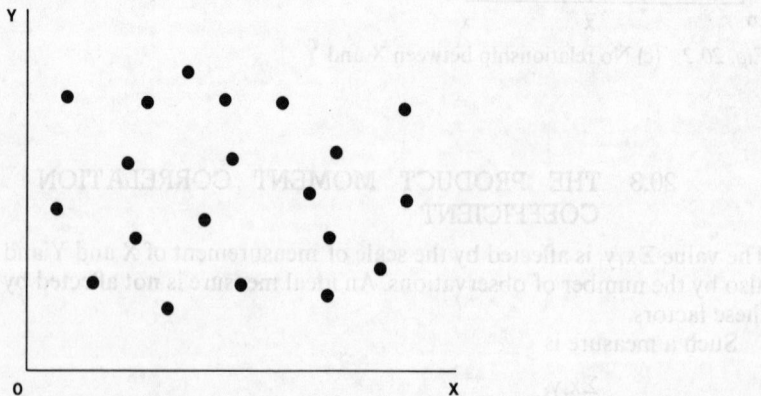

Fig. 20.4 No correlation (r = 0)

Example of positive correlation

If **r = +1** there is perfect positive correlation. A straight line exactly fits the points. High values of X go with high values of Y, as in Fig. 20.5.

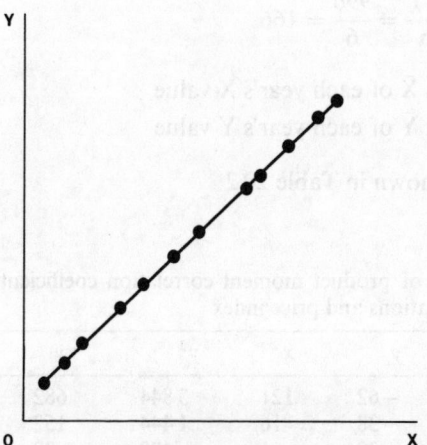

Fig. 20.5 Perfect positive correlation (r = +1)

The closer the value of r to +1, the higher the degree of positive correlation. The closer the value of r to −1, the higher the degree of negative correlation. The closer the value of r to zero, the lower the degree of correlation.

The significance of the value of r is explained in section 20.8.

20.4 EXAMPLE OF POSITIVE CORRELATION

The figures in Table 20.1 relate the number of mortgage applications received by a building society for a particular category of property to an index of property prices for each year. We wish to calculate the product moment correlation coefficient.

Table 20.1 Relation between mortgage applications and price index

Year	Number of applications (X) (00's)	Price index (Y)
1974	30	104
1975	37	128
1976	40	146
1977	42	170
1978	44	218
1979	53	230

The mean of X: $\bar{X} = \dfrac{\Sigma X}{n} = \dfrac{246}{6} = 41$

The mean of Y: $\bar{Y} = \dfrac{\Sigma Y}{n} = \dfrac{996}{6} = 166$

x = the deviation from $\bar{X}$ of each year's X value
y = the deviation from $\bar{Y}$ of each year's Y value

Therefore we have the values shown in Table 20.2

Table 20.2 Calculation of product moment correlation coefficient between mortgage applications and price index

Year	x	y	x^2	y^2	xy
1974	−11	−62	121	3 844	682
1975	−4	−38	16	1 444	152
1976	−1	−20	1	400	20
1977	1	4	1	16	4
1978	3	52	9	2 704	156
1979	12	64	144	4 096	768
			292	12 504	1782

For the correlation coefficient formula, we now have the values

$\Sigma x^2 = 292$
$\Sigma y^2 = 12\,504$
$\Sigma xy = 1782$

$$r = \dfrac{1782}{\sqrt{(292 \times 12\,504)}} = +0.933$$

This value for r indicates a fairly high degree of correlation between the two sets of values.

20.5 EXAMPLE OF NEGATIVE CORRELATION

Table 20.3 shows the index numbers of sales of aluminium and wooden window frames by a certain supplier over a twelve-year period (1972 = 100 in both cases). We wish to find the degree of correlation between these two sets of figures.

Table 20.3 Index numbers of sales of window frames

Year	Index of sales of aluminium frames (X)	Index of sales of wooden frames (Y)
1969	91	116
1970	92	115
1971	91	113
1972	100	100
1973	112	96
1974	119	103
1975	128	96
1976	133	89
1977	139	91
1978	155	92
1979	163	90
1980	165	87
	1488	1188

Finding the means:

$$\bar{X} = \frac{\Sigma X}{n} = \frac{1488}{12} = 124$$

$$\bar{Y} = \frac{\Sigma Y}{n} = \frac{1188}{12} = 99$$

We can then find the values for the correlation coefficient formula. This is shown in Table 20.4.

$$r = \frac{-2825}{\sqrt{(8432 \times 1214)}} = -0.883$$

In this example the correlation is obviously negative and it is important that the negative sign is put in front of the coefficient to show this.

20.6 RANK CORRELATION

Sometimes it is necessary to know whether there is any association between two variables, without necessarily needing to know the extent of the association.

Such a measure can be more easily computed than the product moment coefficient.

303

Correlation

Table 20.4 Calculation of product moment correlation coefficient between sales of aluminium and wooden window frames

Year	x	y	x^2	y^2	xy
1969	−33	17	1089	289	−561
1970	−32	16	1024	256	−512
1971	−33	14	1089	196	−462
1972	−24	1	576	1	−24
1973	−12	−3	144	9	+36
1974	−5	4	25	16	−20
1975	4	−3	16	9	−12
1976	9	−10	81	100	−90
1977	15	−8	225	64	−120
1978	31	−7	961	49	−217
1979	39	−9	1521	81	−351
1980	41	−12	1681	144	−492
			8432	1214	−2825

20.7 SPEARMAN'S COEFFICIENT OF RANK CORRELATION

This measure is calculated from data which have been put into ranked descending or ascending order.

The measure denoted by r_s is found by the following formula:

$$r_s = 1 - \frac{6\Sigma d^2}{n(n^2 - 1)}$$

where d = the difference between the X and Y rankings for any pair of values.

n = the number of pairs of values.

The values obtained for r_s can vary between 0 and ±1. It is important that the two sets of values are ranked in the same way to obtain the correct sign for the coefficient.

Example 20.1

The examination marks which ten students obtained in Surveying and Statistics are given in Table 20.5. We wish to find whether there seems to be any association between the two sets of marks.

If the students' marks are ranked in descending order then the difference in rankings between the two subjects can be measured as in Table 20.6.

n = 10
$\Sigma d^2 = 24$

Spearman's coefficient of rank correlation

Table 20.5 Examination marks of ten students

Student	Surveying	Statistics
A	53	67
B	68	59
C	58	68
D	42	58
E	50	40
F	92	85
G	41	57
H	94	69
I	23	46
J	29	23

Table 20.6 Difference in rankings of marks for two subjects

Student	Surveying (X)	Statistics (Y)	Rank of X	Rank of Y	d	d^2
A	53	67	5	4	1	1
B	68	59	3	5	−2	4
C	58	68	4	3	1	1
D	42	58	7	6	1	1
E	50	40	6	9	−3	9
F	92	85	2	1	1	1
G	41	57	8	7	1	1
H	94	69	1	2	−1	1
I	23	46	10	8	2	4
J	29	23	9	10	−1	1
						24

and $\quad r_s = 1 - \dfrac{6 \times 24}{10(100 - 1)} = 1 - \dfrac{144}{990} = +0.85$

The main problem of calculation of the rank correlation coefficient arises when two or more values are ranked equally.

There are two ways to deal with this problem.

1. If two values are, say, ranked equal 2nd, then give them both a ranking of 2 and rank the next value as 4th.
2. Alternatively, the rankings could be shared, eg in the above mentioned case the values ranked 2nd could be considered to be sharing 2nd and 3rd rankings and so could both be given an averaged ranking of $(2 + 3)/2 = 2\frac{1}{2}$.

As before, the next value would be ranked 4th. Whichever method is chosen it is important that it is consistently pursued throughout the calculation.

Correlation

20.8 THE INTERPRETATION OF THE VALUE OF r

Obviously the closer to unity ($+1$ or -1) is the value of the coefficient, the higher is the degree of correlation, and a value close to 0 indicates that little correlation exists.

The extreme values of 1 or 0 are unlikely to occur very often in practice though and in most examples we are usually concerned with interpreting the significance of values between perhaps 0.5 and 0.99.

The value of r is affected by the numbers of pairs of values for the two variables. For a small sample of items, the value of r obtained should be treated carefully. With a small number some degree of correlation may arise purely by chance.

For a larger sample, a test of significance for r involves the calculation of the standard error of r which is written as

$$\sigma_r = \frac{1}{\sqrt{(n-1)}}$$

If the calculated r is more than approximately twice (or 1.96 to be precise) this standard error then we can say at the 0.05 level of confidence that there is a significant correlation.

Perhaps the most common misuse of the correlation coefficient arises when it is used to imply a causal relationship between variables. This is not its purpose. It does nothing more than show the strength of a linear relationship.

A correlation coefficient calculated for two variables may have a high value, but the correlation may be of a spurious nature, ie the two variables may not be causally related but may both be associated with some other outside factor.

20.9 THE CORRELATION COEFFICIENT FORMULA: A MATHEMATICAL NOTE

The formula for the correlation coefficient was given in section 20.3 as

$$r = \frac{\Sigma x_i y_i}{\sqrt{(\Sigma x_i^2 \Sigma y_i^2)}}$$

as this is a form which is considered to be the simplest to understand and to utilise.

However, the formula may be rewritten as

$$r = \frac{n\Sigma X_i Y_i - (\Sigma X_i)(\Sigma Y_i)}{\sqrt{([n\Sigma X_i^2 - (\Sigma X_i)^2][n\Sigma Y_i^2 - (\Sigma Y_i)^2])}}$$

This form has the virtue that it is not necessary to find the mean values and the deviations from the means as the data in its original form is used. This may make for easier calculation in some examples.

To show the equivalence of these two versions of the correlation coefficient:

We defined $x_i = X_i - \bar{X}$ and $y_i = Y_i - \bar{Y}$

$$\therefore \quad r = \frac{\Sigma x_i y_i}{\sqrt{(\Sigma x_i^2 \Sigma y_i^2)}}$$

$$= \frac{\Sigma(X_i - \bar{X})(Y_i - \bar{Y})}{\sqrt{(\Sigma(X_i - \bar{X})^2 \Sigma(Y_i - \bar{Y})^2)}}$$

Consider $\Sigma(X_i - \bar{X})(Y_i - \bar{Y})$ when expanded

$$= \Sigma(X_i Y_i - X_i \bar{Y} - \bar{X} Y_i + \bar{X}\bar{Y})$$
$$= \Sigma X_i Y_i - \bar{Y}\Sigma X_i - \bar{X}\Sigma Y_i + n\bar{X}\bar{Y}$$
$$= \Sigma X_i Y_i - \frac{\Sigma Y_i}{n}\Sigma X_i - \frac{\Sigma X_i}{n}\Sigma Y_i + n\frac{\Sigma X_i}{n}\frac{\Sigma Y_i}{n}$$
$$= \Sigma X_i Y_i - \frac{(\Sigma X_i)(\Sigma Y_i)}{n}$$
$$= \frac{n\Sigma X_i Y_i - (\Sigma X_i)(\Sigma Y_i)}{n}$$

In a similar fashion,

$$\Sigma(X_i - \bar{X})^2 = \frac{n\Sigma X_i^2 - (\Sigma X_i)^2}{n}$$

$$\Sigma(Y_i - \bar{Y})^2 = \frac{n\Sigma Y_i^2 - (\Sigma Y_i)^2}{n}$$

$$r = \frac{\dfrac{n\Sigma X_i Y_i - (\Sigma X_i)(\Sigma Y_i)}{n}}{\sqrt{\left(\left[\dfrac{n\Sigma X_i^2 - (\Sigma X_i)^2}{n}\right]\left[\dfrac{n\Sigma Y_i^2 - (\Sigma Y_i)^2}{n}\right]\right)}}$$

$$= \frac{n\Sigma X_i Y_i - (\Sigma X_i)(\Sigma Y_i)}{\sqrt{([n\Sigma X_i^2 - (\Sigma X_i)^2][n\Sigma Y_i^2 - (\Sigma Y_i)^2])}}$$

20.10 EXERCISES

1. The data contained in Table 20.7 can be used to assess the strength of the link between yields on short-term and long-term investments. Use the set of data in the table to demonstrate how the product moment

Correlation

correlation coefficient can be estimated. Indicate how these results could be improved (further calculations are not required).

Table 20.7 Short and long-term yields

Year	At last day of month	Three month treasury bill yield (%)	2½% Consols Yield (%)
1978	January	5.83	10.02
	February	6.25	11.30
	March	6.02	11.42
	April	7.03	12.01
	May	8.55	12.20
	June	9.33	12.41
	July	9.19	12.23
	August	8.91	12.28
	September	9.19	12.33
	October	10.36	12.46
	November	11.56	12.51
	December	11.64	12.33

(*Sources:* Financial Statistics and *Estates Gazette*) (RICS)

2. Use the data contained in the following table to estimate the level of linear association between the two factors. Comment on the adequacy of this statistical measure in this specific context.

Region	Average per cent growth in land prices (1969–73)	Per cent growth in population (1971–72)
Wales	16.2	0.33
Northern	36.3	0.09
East Anglia	41.5	1.43
East Midlands	29.4	0.72
Yorkshire & Humberside	23.8	0.21
West Midlands	21.9	0.53
South West	34.0	1.08
North West	28.0	0.06
South East	30.7	0.38
England & Wales	30.0	0.32

(RICS)

3. (a) Calculate the Pearson product-moment correlation coefficient for the following data.

	Office moves. No. of firms (1963–79) 00	Aggregate rateable values (1979) £ 0m
Areas		
East Anglia	0.5	0.6
South East	18.1	47.8
Greater London	8.1	43.2
Rest of South East	2.6	4.6
West Midlands	0.5	1.6
Wales	0.2	0.5

(*Sources*: Inland Revenue and Department of the Environment)
(b) To what extent can the data and your coefficient, on the face of them, without further information can be taken as evidence that in the UK high rates are driving out offices? (RICS)

21 TIME SERIES ANALYSIS

21.1 INTRODUCTION

A time series is the successive measurement of the size or value of a variable over time.

Example 21.1

A typical example of how data vary over time is shown by example in Table 21.1, which shows the number of permanent dwellings started in Northern England for each quarter during the period 1977–79.

Table 21.1 Permanent dwellings starts in each quarter (Source: *Housing and Construction Statistics*, No. 32, Government Statistical Service)

Year	Quarter	Starts
1977	1st quarter	2922
	2nd quarter	5197
	3rd quarter	3975
	4th quarter	3203
1978	1st quarter	2284
	2nd quarter	4713
	3rd quarter	3679
	4th quarter	3082
1979	1st quarter	1626
	2nd quarter	4273
	3rd quarter	3060
	4th quarter	3380

Many sets of figures which vary over time have an element of **seasonal** or **cyclical** variation in their pattern, eg house building statistics often show a marked variation between the summer and winter months, property sales statistics indicate that sales move more rapidly in periods of economic growth.

In many cases it may be useful to remove the variation element from a time series in order to estimate the **trend** for prediction purposes.

21.2 THE METHOD OF MOVING AVERAGES

We can deal with seasonal variations by the **method of moving averages**. From the actual time series data in the example, we are interested in removing the seasonal element (eg the fact that the number of housing starts is invariably higher in the second quarter than in the first quarter) in order to find the **trend**.

The steps we take to employ this method are as follows:

(a) total the values four at a time, each time dropping one value and adding on the next one, and set the total against the second value;
(b) add the four quarter totals, two at a time, to give the centred totals (centring is necessary whenever there is an even number of items in the moving average);
(c) divide the values in the centred tables by 8 to find the **trend**. The reason for this is that each of the values in this centred table column is made up of eight values from the time series column.

If we take Example 21.1, the trend is found in Table 21.2. As can be seen from the resultant trend, the seasonal variations present in the original time series have now been removed. The values in the seasonally adjusted trend column show the downward movement in the series once the seasonal variation has been removed.

Table 21.2 Calculation of trend of housing starts

Year	Quarter	Housing starts	Total four quarters	Centred table	Trend
1977	I	2922	–	–	–
	II	5197	15 297	–	–
	III	3975	14 659	29 956	3 744.5
	IV	3203	14 175	28 834	3 604.25
1978	I	2284	13 879	28 054	3 506.75
	II	4713	13 758	27 637	3 454.625
	III	3679	13 100	26 858	3 357.25
	IV	3082	12 660	25 760	3 220.0
1979	I	1626	12 041	24 701	3 087.625
	II	4273	12 339	24 380	3 047.5
	III	3060	–	–	–
	IV	3380	–	–	–

Time series analysis

One unfortunate feature of this method, however, is that the number of values in the trend column is inevitably less than the number in the original time series; values at the beginning and end of the series being 'chopped off'.

The original time series data and the trend line are shown in Fig. 21.1. The smoothing effect of this method of moving averages is clearly shown.

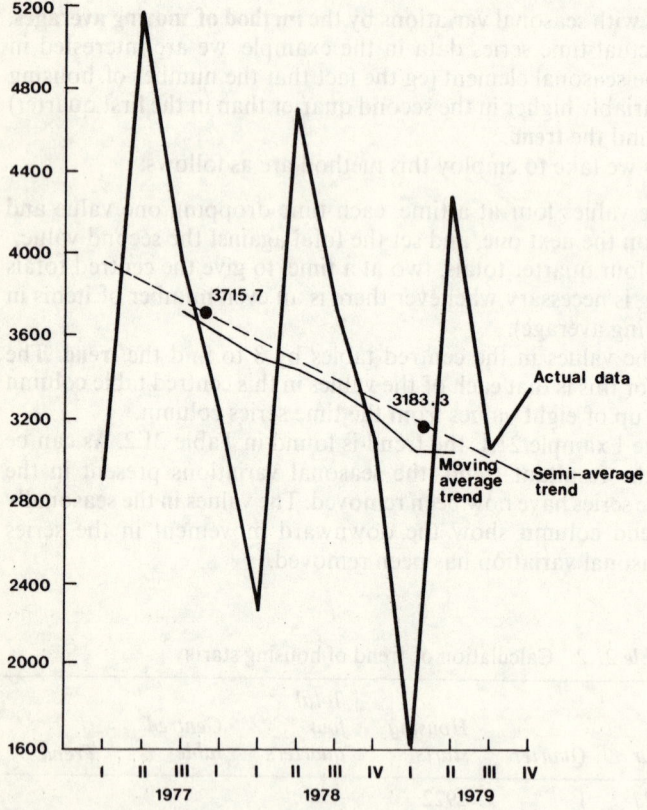

Fig. 21.1 Original time series, moving average trend and semi-average trend: Example 21.1

21.3 CYCLICAL VARIATIONS

Due to the influence of booms and slumps in the economy, construction activity often follows a wave-like pattern over the years. In practice this periodic cycle may last between five and fifteen years; this period being measured from one peak to the next.

From a time series affected by this type of cycle the variation due to

Cyclical variations

this element may again be removed by the method of moving averages.

Example 21.2

Table 21.3 shows the turnover of a construction company for the previous fifteen years (year 15 is the most recent). From the figures the nature of cyclical pattern may not be immediately apparent. When the figures are plotted graphically, however, a five-year pattern can be perceived.

Table 21.3 Turnover of a construction company

Year	Turnover (£m)
1	36
2	33
3	43
4	46
5	38
6	43
7	40
8	48
9	52
10	40
11	48
12	43
13	53
14	57
15	47

The cycle is twin peaked with a mini-boom in years 1, 6 and 11 and the major peak of the cycle also occurs at five-yearly intervals, ie years 4, 9 and 14. In the construction industry this type of pattern is not uncommon.

To eliminate the cyclical variation from this series, the three five-year cycles can be identified, as too can the 1st period, 2nd period, etc of each cycle. So, for instance, year 5 becomes 1st cycle, 5th period; year 6 becomes 2nd cycle, 1st period and so on.

A five-yearly moving average can then be taken in order to arrive at the trend. This is carried out in Table 21.4. The calculation of the trend in this case is simpler than in the previous example due to the fact that there is no need to draw up centred totals. The totals of five values can be placed directly against the centre of the set of values, ie the first five values sum to 196 and this figure is put against the middle of these five values the 43.

The values in the totals column can then be directly divided by 5 to give the trend figures.

The original time series and smoothed trend are shown in Fig. 21.2.

Time series analysis

Table 21.4 Calculation of five-yearly moving average

Year	Cycle	Period of cycle	Turnover (£m)	Five-year total	Trend
1	1st	1st	36	–	–
2		2nd	33	–	–
3		3rd	43	196	39.2
4		4th	46	203	40.6
5		5th	38	210	42.0
6	2nd	1st	43	215	43.0
7		2nd	40	221	44.2
8		3rd	48	223	44.6
9		4th	52	228	45.6
10		5th	40	231	46.2
11	3rd	1st	48	236	47.2
12		2nd	43	241	48.2
13		3rd	53	248	49.6
14		4th	57	–	–
15		5th	47	–	–

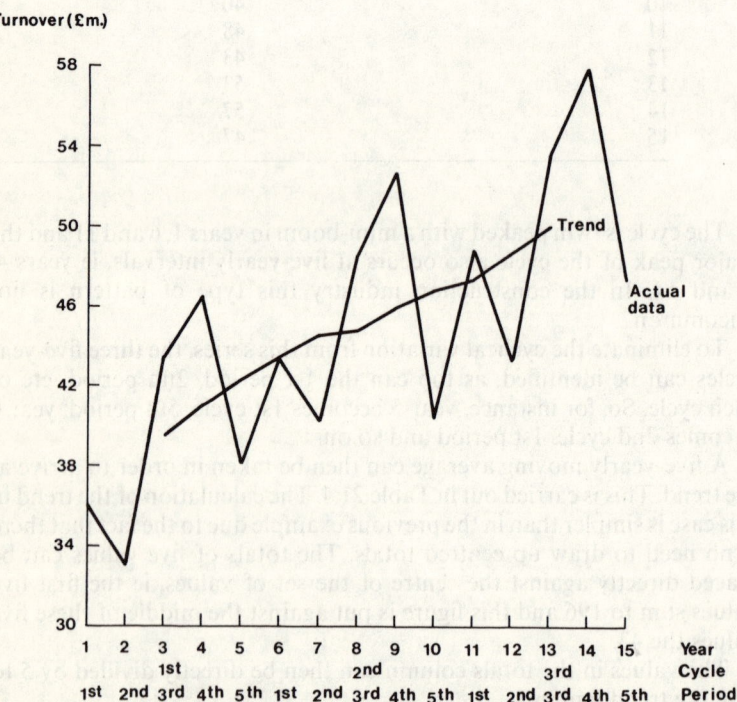

Fig. 21.2 Original time series and estimated trend: Example 21.2

This method of moving averages can be used for any time series where a regular pattern exists. A three-year moving average may be appropriate in certain series. Certainly twelve-period moving averages are commonly encountered in the case of monthly statistics as are seven-period moving averages for daily produced data.

The major qualification to the use of the moving average method is that the trend inevitably covers a shorter period than the original time series and so a calculation involving a long moving average period must use a correspondingly large amount of data; enough to identify the regularity in the pattern.

21.4 THE USE OF TIME SERIES ANALYSIS

The main purpose of this type of analysis is to find out more about the behaviour of a time series by separating the various seasonal variations and/or cyclical variations from the trend. In forward planning a knowledge of the trend contained in a series is vital and the identification of any pattern or regularity can greatly assist any prediction of future values in the series.

21.4.1 Extrapolation

In Example 21.2 the company in question may well be interested in estimating the turnover in years 16 and 17. To undertake such a calculation two factors would have to be considered.

The trend in the series. From the trend figures it can be seen that the trend rose from 39.2 to 49.6 between year 3 and year 13. This is a rise of 10.4 over the 10 years and thus constitutes an average increase of 1.04.

Using this average as the estimate of the annual increase in the trend, then three years after the last trend value, ie year 16, we can expect the trend to have risen to year 13's value + (3 × annual average increase) = 49.6 + (3 × 1.04) = 52.72.

In a similar fashion the estimate for the trend value for year 17 is 49.6 + (4 × 1.04) = 53.76.

The cyclical factor. Our estimate for a particular year must take account of the particular point in a cycle in which a year falls. In the designation of the five-year pattern year 16 is the 1st period of the 4th cycle and year 17 is the 2nd period of the 4th cycle.

The simplest way to take account of the cyclical factor is to find the average variation from the trend of all the 1st period years, 2nd period years, etc for which we have the data and to adjust our predicted trend values by the appropriate average period of cycle variation. How this is done is shown is Tables 21.5 and 21.6.

Time series analysis

Table 21.5 Calculations of variations from the trend

Cycle	Period	Turnover (£m)	Trend	Variation from trend
1st	1st	36	–	
	2	33	–	
	3	43	39.2	+3.8
	4	46	40.6	+5.4
	5	38	42.0	−4.0
2nd	1	43	43.0	0.0
	2	40	44.2	−4.2
	3	48	44.6	+3.4
	4	52	45.6	+6.4
	5	40	46.2	−6.2
3rd	1	48	47.2	+0.8
	2	43	48.2	−5.2
	3	53	49.6	+3.4
	4	57	–	–
	5	47	–	–

Table 21.6 Calculation of average variation for each year of the cycle

	1st period	2nd period	3rd period	4th period	5th period
1st cycle	–	–	+3.8	+5.4	−4.0
2nd cycle	0.0	−4.2	+3.4	+6.4	−6.2
3rd cycle	+0.8	−5.2	+3.4	–	–
Total	+0.8	−9.4	+10.6	+11.8	−10.2
Average	+0.4	−4.7	+3.53	+5.9	−5.1

The figures in the variation from the trend column in Table 21.5 are found by subtracting the trend figure from the original time series figure for each year. So if the trend figure is less/more than the corresponding time series figure the variation has a positive/negative sign.

The variation figures are then set out in rows for each cycle as in Table 21.6 in order to find the average variation for each particular period of the cycle. (Note that the 3rd period's column total is the only one formed from three values because the first and last years in the trend column happen to fall at this point in the cycle. This column total must therefore be divided by three, unlike all the others which are divided by only two.)

To return to the problem in hand, the trend figures we have found for years 16 and 17 must now be adjusted to allow for seasonal variations. Year 16's trend estimate of 52.72 needs to have +0.4 added to it as this is our estimate of the average 1st period of the cycle variation. This gives a total of 53.12. Year 17's trend estimate of 53.76 must similarly be adjusted by the average 2nd period variation of -4.7. This gives a total of 49.06.

Our final estimates, therefore, of the company's turnover in years 16 and 17 are £53.12 m. and £49.06 m. respectively.

21.5 THE SEMI-AVERAGE METHOD

As the simplest form of trend is the straight line, one basic way of graphical estimation is to draw a straight line through the data to represent the trend. One method of doing this is the semi-average method.

This involves splitting the time series chronologically into halves and finding the mean of the figures in each half. The mean value is then plotted against the centre in each half and these two points are then joined together.

As an example, the data on housing starts is split in this way in Table 21.7 and then the two means are found. The line formed by joining these points is shown in Fig. 21.1, together with the estimated trend line from the moving average method and the actual data.

The acceptability of this simple method depends very much upon whether it seems reasonable to fit a straight line to the data.

Table 21.7 Calculation of semi-average for housing starts

2 922	3 679
5 197	3 082
3 975	1 626
3 203	4 273
2 284	3 060
4 713	3 380
$\dfrac{22\,294}{6} = 3\,715.7$	$\dfrac{19\,100}{6} = 3\,183.3$

21.6 TIME SERIES AND THE LEAST SQUARES METHOD

As we have seen in Ch. 19, a useful way of fitting a straight line to two sets of data is the least squares method.

We can apply this method to time series data when we can assume that the value of the variable is determined by the time period involved.

Time series analysis

The two normal equations can be used to produce a linear equation $Y = a + bX$, where time is the independent variable, and where

a = the value of the variable in period zero (ie normally the starting value in the series)

X = the number of periods beyond this starting point.

Example 21.3

The data in Table 21.8 show the mean value of all the residential properties sold by an estate agency during a nine year period. We can find the equation of the trend line in the same manner as was used to find the regression line in earlier examples. However, due to the normally regular pattern (ie successive years in this example) of time series data, the procedure can be greatly simplified by changing the scale of the X's so that their sum is equal to zero.

Table 21.8 Mean value of residential properties sold by an estate agency

Year	1971	1972	1973	1974	1975	1976	1977	1978	1979
Value (£)	4300	6100	7100	8200	9400	11 500	14 600	17 300	19 500

When this is done the normal equations

$$\Sigma Y = na + b\Sigma X$$

and

$$\Sigma XY = a\Sigma X + b\Sigma X^2$$

become

$$\Sigma Y = na + (b \times 0)$$
$$\therefore \quad \Sigma Y = na$$
$$\therefore \quad a = \frac{\Sigma Y}{n}$$

and

$$\Sigma XY = (a \times 0) + b\Sigma X^2$$
$$\therefore \quad \Sigma XY = b\Sigma X^2$$
$$\therefore \quad b = \frac{\Sigma XY}{\Sigma X^2}$$

This simplification is achieved by setting the origin in the middle of the time periods. In the case of an odd number of periods the middle period becomes 0 but in the case of an even number the origin falls

Time series and the least squares method

between the middle two, so in this case we change the time scale to read:

```
1971  1972  1973  1974  1975  1976  1977  1978  1979
 -4    -3    -2    -1     0     1     2     3     4
```

The values necessary for a and b can now be found from Table 21.9.

Table 21.9 Change of time scale and calculation of a and b to find regression equation

	X	Y	X^2	XY
1971	-4	4 300	16	-17 200
1972	-3	6 100	9	-18 300
1973	-2	7 100	4	-14 200
1974	-1	8 200	1	-8 200
1975	0	9 400	0	0
1976	1	11 500	1	11 500
1977	2	14 600	4	29 200
1978	3	17 300	9	51 900
1979	4	19 500	16	78 000
	0	98 000	60	112 700

$\Sigma Y = 98\,000$, $\Sigma X^2 = 60$, $\Sigma XY = 112\,700$, $n = 9$

$$a = \frac{\Sigma Y}{n} = \frac{98\,000}{9} = 10\,888.9$$

$$b = \frac{\Sigma XY}{\Sigma X^2} = \frac{112\,700}{60} = 1878.3$$

∴ The regression equation is

$Y = 10\,888.9 + 1878.3\,X$

It must be made clear that this is the trend line which gives an estimated value of £10 888.9 for 1975 and not for 1971, the first year in the series.

This least squares trend line together with the moving average trend and actual data is shown in Fig. 21.3.

The estimated value for 1971 is found by putting an X value of -4 into the regression equation, ie

$$Y = 10\,888.9 + (1878.3 \times -4)$$
$$= 10\,888.9 - 7513.3$$
$$= £3375.6$$

319

Time series analysis

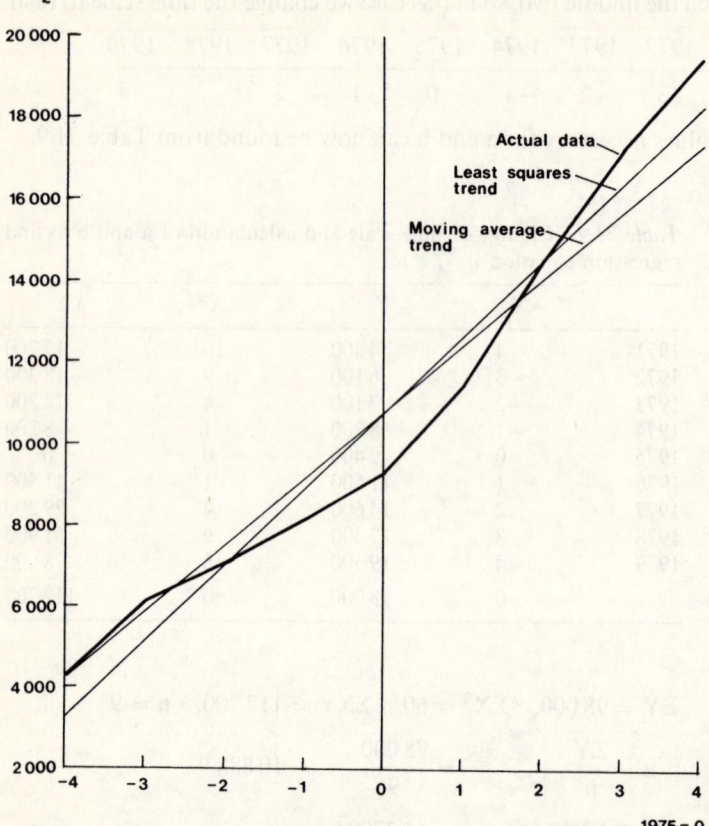

Fig. 21.3 Original time series, moving average trend and least squares trend: Example 21.3

We can carry out extrapolation in the normal manner for the regression equation. So that if we want to estimate the value for 1980, for instance, we put in an X value of $+5$ into the equation, ie

$$Y = 10\,888.9 + (1878.3 \times 5)$$
$$= 10\,888.9 + 9391.5$$
$$= \underline{£20\,280.4}$$

21.7 EXERCISES

1. Define a time series and give three examples of such a series. The following data give the construction index for a firm for a nine-year period.

Year 1964 1965 1966 1967 1968 1969 1970 1971 1972
Index 117 108 102 129 123 111 141 132 123

Examine the data and smooth the data using an appropriate moving average. Comment on the trends in the above data. (RICS)

2. Describe how you would calculate a four-point centred moving average. The following data give the amount of new projects started in £ million.

Year	Quarter			
	I	II	III	IV
1971	148(—)	20(—)	32(62)	40(65)
1972	164(67)	28(70)	40(74)	56(77)
1973	180(79)	36(81)	48(—)	64(—)

The second figure for each quarter is the four-point moving average. Calculate the average seasonal deviation for each quarter.

If the value for the first quarter of 1974 is 214 and, assuming that the seasonal variations are average, give an estimate for the three remaining quarters of 1974. (RICS)

3. (a) Explain what is meant by:
   (i) time series and give examples which relate to the work of a surveyor;
   (ii) components of a time series;
   (iii) time series analysis and show how it may be carried out in order to isolate the components.

(b) Briefly discuss the ways in which a surveyor might utilise the results of such an analysis and indicate its possible benefits and limitations. (RICS)

22 SAMPLING METHODS AND SAMPLING DISTRIBUTIONS

22.1 INTRODUCTION

Sampling from a population represents an attempt to assess some properties of a large number of items by a study of similar properties of a smaller group. Sampling is useful in a number of situations.

1. Where the whole of the population is not easily accessible, so that sampling is the only means available (eg taking sample bores on site to ascertain geological structure).
2. Where a study of the whole population would be too expensive or take too long (eg collecting information on price changes from all retail outlets for a price index).
3. Where the attribute being sampled would involve the destruction of the item (eg testing the breaking strength of building materials).

The usefulness of sampling depends upon two factors:

(a) the sample must be chosen from the population in such a way that each member of the population has an equal chance of being chosen. This is known as random sampling;
(b) the size of the sample is also important. Generally the larger the sample the more reliance can be placed on the results being a cross-section of the population.

22.2 SAMPLING METHODS

Random sampling. If a sample is chosen in a purely random manner a representative selection can be obtained but this can be quite a difficult process. It may be impossible to take a perfectly random sample in many cases but the avoidance of bias is extremely important.

Sampling methods

The best method of random sampling is to use a table of random numbers. Such a table is so called because there is no bias in the sequence of digits. While the digits may be arranged as in Table 22.1, this has no significance. The table may be read up or down and the numbers used, one, two, three, etc, at a time as required.

Table 22.1 Random numbers

7 2	9 5	1 4	0 4	8 4
9 7	4 2	7 2	3 6	0 8
8 0	2 1	0 1	4 9	6 2
1 5	5 7	3 5	0 3	4 6
6 7	2 4	6 3	7 1	3 4
8 0	1 6	9 5	6 3	6 8
1 7	9 1	5 9	9 5	3 2
7 9	3 5	1 3	8 4	6 7
0 6	8 1	2 7	0 5	2 9
0 8	3 4	2 8	9 0	5 8

Each digit is an independent sample taken from a population in which the digits 0, 1, ... 9 are equally likely to occur.

Different methods of applying random sampling may be used, depending on the kind of population one wishes to survey. Such methods are considered below.

Stratified sampling. This approach can be used when there are obvious sub-groups within the population and it is desirable to make sure that each sub-group is suitably represented in the sample.

For instance, it may be known that 70 per cent of the private construction firms in a region are small firms (employing less than twenty workers), 20 per cent are medium sized (between twenty one and 100 workers) and 10 per cent are large (over 100 workers). So if a sample of 200 firms were chosen for a survey and the sample were stratified according to size, the surveyed firms would be apportioned as follows: 140 small firms:40 medium-sized firms:20 large firms. This would be called a stratified sample. A fair representation is assured.

The attributes being surveyed can then be compared for the various groups.

Systematic sampling. This is a refinement of the basic method of simple random sampling. The method is to begin from a randomly selected item and then take every nth item (n being chosen to give the number required in the sample).

For example, there may be 500 houses on an estate and a sample survey of fifty houses may be required. One of the first ten houses could be chosen at random and then every tenth house thereafter.

Such a method is more convenient than ordinary random sampling and can be just as effective.

Multi-stage sampling. This method involves the reduction of a population to the required sample size by making choices from the population at various stages and the use of random sampling to make these choices at each stage.

If it were desired to interview 1000 people at random from the various regions of England, ten counties could be chosen at random initially; within each county two districts could then be picked, and within each district a random sample of fifty people selected. This would give the total sample of 1000.

The main advantage of this method is that the actual sampling procedure is likely to be made easier due to the greater concentration of the sample than would occur in a purely random sampling procedure from the total population.

The size of the sample. The larger the size of a sample, the more likely it is to give a representative picture of the population from which it has been chosen.

The accuracy of estimates of the characteristics of a population devised from sample information depends upon the size of the sample. As a general rule, the size of a sample should be at least thirty.

The problem in many sampling situations though is that the larger the sample taken the greater the expense, time and effort involved and these factors have to be balanced against the extra accuracy which may result from taking a larger sample.

22.3 SAMPLING DISTRIBUTIONS

One main purpose of taking a sample from a population is to enable an estimate to be made of the mean of that population on the basis of information from the sample.

We are therefore interested in knowing what the probabilities are that a sample mean is likely to differ from the population mean by certain amounts.

If all the possible samples of size n were taken from a population and the mean of each sample taken, then a sampling distribution of these means could be formed.

The mean of such a distribution of the sample means is called $\mu_{\bar{x}}$ and the standard deviation of the sample means from the true population mean is called $\sigma_{\bar{x}}$ and is referred to as the standard error of the mean.

The larger the size of the samples taken, the smaller the difference between a sample mean and the population mean can be expected to be.

So the standard error is therefore given by the formula:

$$\sigma_{\bar{x}} = \sigma/\sqrt{n}$$

where σ = the standard deviation of the population; which can be substituted by the standard deviation of the sample (s) if the sample is large enough, ie

$$n \geqslant 30$$

Also if $n \geqslant 30$, the Central Limit theorem is applicable. This states that the theoretical sampling distribution of $\bar{x}$ can be approximated to the normal distribution regardless of the way in which the population is distributed.

Using this theorem it is possible to calculate the probability of obtaining various values for $\bar{x}$ (the sample mean).

Example 22.1

If the true mean height of all the students in a large college is 174.5 cm and the standard error is 4.5 cm, find the probability that a sample of 100 students will have a mean height of more than 175.2 cm.

The mean of the sampling distribution is equal to the population mean (μ)

$$\mu_{\bar{x}} = \mu = 174.5$$

and the standard error of the sampling distribution is

$$\sigma_{\bar{x}} = \sigma/\sqrt{n} = \frac{4.5}{10} = 0.45$$

The standard normal distribution for x and the calculation of z values were explained in section 17.9. In a similar manner, the sampling distribution of $\bar{x}$, with mean $\mu_{\bar{x}}$ and standard error $\sigma_{\bar{x}}$ can also be standardised. We have:

$$z = \frac{\bar{x} - \mu_{\bar{x}}}{\sigma_{\bar{x}}}$$

Putting the appropriate values into the formula gives:

$$z = \frac{175.2 - 174.5}{0.45} = 1.56$$

We are interested in finding the area to the right of $z = 1.56$ as shown in Fig. 22.1.

From the z table, the area to the left of $z = 1.56$ is 0.9406 and so the area we want is $1 - 0.9406 = 0.0594$, which is the probability of the sample having a mean greater than 175.2 cm.

Sampling methods and sampling distributions

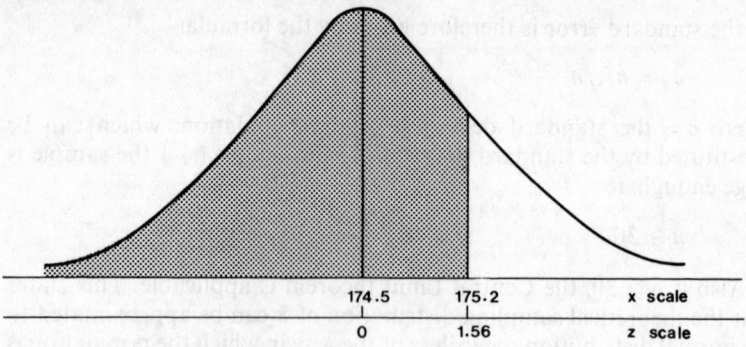

Fig. 22.1 Conversion to the z scale

22.4 THE USE OF CONFIDENCE LIMITS IN THE ESTIMATION OF MEANS

If a single sample is taken and the sample mean found then it is obviously impossible to determine the population mean precisely from such information. What we can do, though, is to establish limits within which the true population mean will fall, with a specified probability.

These limits are called confidence limits, and their determination is based upon our knowledge of the characteristics of the normal distribution.

As already shown in section 17.8, we know that 95 per cent of the area under a normal curve lies within two (or 1.96 to be exact) standard deviations of the mean value.

So, for the sampling distribution of means which has a mean $\mu_{\bar{x}}$ and standard error of $\sigma/\sqrt{n}$ we can say that if $\bar{x}$ is the mean of a sample, then with a probability of 0.95 the standardised z value $(= (\bar{x} - \mu_{\bar{x}})/(\sigma/\sqrt{n}))$ will lie between -1.96 and $+1.96$.

This can be written as:

$$-1.96 < \frac{\bar{x} - \mu_{\bar{x}}}{\sigma/\sqrt{n}} < +1.96$$

$$\therefore \quad -1.96 \frac{\sigma}{\sqrt{n}} < \bar{x} - \mu_{\bar{x}} < +1.96 \frac{\sigma}{\sqrt{n}}$$

$$\therefore \quad \bar{x} - 1.96 \frac{\sigma}{\sqrt{n}} < \mu_{\bar{x}} < \bar{x} + 1.96 \frac{\sigma}{\sqrt{n}}$$

Thus with a probability of 0.95, ie 95 per cent confidence, we can claim that the interval $\bar{x} - 1.96(\sigma/\sqrt{n})$ to $\bar{x} + 1.96(\sigma/\sqrt{n})$ contains $\mu_{\bar{x}}(=\mu)$. This is therefore called the 95 per cent confidence interval.

Significance tests

As stated earlier, the mean of the sampling distribution is equal to the population mean and for a large sample σ is approximated by the sample standard deviation, s.

This means that the confidence interval can be rewritten as

$$\bar{x} - 1.96\frac{s}{\sqrt{n}} < \mu < \bar{x} + 1.96\frac{s}{\sqrt{n}}$$

Example 22.2

A random sample of the weekly wages of 900 men employed in the construction industry in 1979 gave a mean weekly wage of £95, with a standard deviation of £8.40.

With 95 per cent confidence, find the limits within which the population mean weekly wage lies.

Here $\bar{x} = 95$, $s = 8.4$ and $n = 900$. So the 95 per cent confidence interval is

$$95 - 1.96\left(\frac{8.4}{\sqrt{900}}\right) < \mu < 95 + 1.96\left(\frac{8.4}{\sqrt{900}}\right)$$

$$89.51 < \mu < 100.49$$

ie the 95 per cent confidence limits are £89.51 and £100.49.

It may well be the case that in certain situations it is necessary to find an interval within which our estimate of the population mean lies with a greater degree of confidence.

We can therefore further specify 98 per cent confidence limits as:

$$\bar{x} \pm 2.33\frac{s}{\sqrt{n}}$$

and 99 per cent confidence limits as:

$$\bar{x} \pm 2.58\frac{s}{\sqrt{n}}$$

Obviously in any particular estimation the greater the degree of confidence the wider apart the limits must be.

When sample information is used as a basis for the estimation of the population mean then the confidence interval needs to be specified.

22.5 SIGNIFICANCE TESTS

Sometimes when sample information is used or experiments are carried out we may wish to find out whether the results obtained are within the range within which we expect them to be.

What often happens is that a hypothesis is put forward (the null hypothesis) and we wish to test whether there is consistency between the sample results and the hypothesis.

A test of significance can be used to 'test' the hypothesis. On the basis of sample information it is possible that we could reject the null

Sampling methods and sampling distributions

hypothesis when the hypothesis is in fact a true one. We can specify the probability of such an error and call it the level of significance. We usually use 0.05 or 0.01.

Example of a significance test. Suppose we want to test the hypothesis that the average value of a type of property in a certain area is £20 000 and we find that a random sample of 100 properties in the area had a mean value of £19 500 and a standard deviation of £2400.

Our null hypothesis (H_0) can be stated as

$$H_0 : \mu = 20\,000$$

and as the alternative is that the population mean value is not £20 000 we have an alternative hypothesis (H_1) of

$$H_1 : \mu \neq £20\,000$$

We specify a level of significance of 0.05. As we know that $z = (\bar{x} - \mu)/(s/\sqrt{n})$ has a standard normal distribution, we can reject the null hypothesis if z is outside the range -1.96 to $+1.96$ (the critical values) but accept it if it is within that range, as shown in Fig. 22.2. This is called the decision rule. We have here

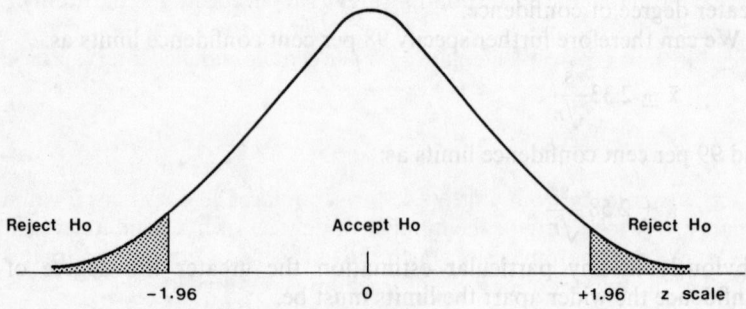

Fig. 22.2 Two-tail significance test (0.05 level)

$$z = \frac{19\,500 - 20\,000}{2400/\sqrt{100}} = -2.08$$

This means we reject the hypothesis that the population mean value is £20 000 at the 0.05 level of significance. Significance tests at the 0.01 level can also be carried out using critical values of ± 2.575. Also 'one-tail' tests using a critical value of $+1.645$ (or -1.645) at the 0.05 level (as depicted in Fig. 22.3) can be used instead of the normal 'two tail test' if we wish to stipulate an alternative hypothesis that the population mean is above (or below) a certain level.

Small samples

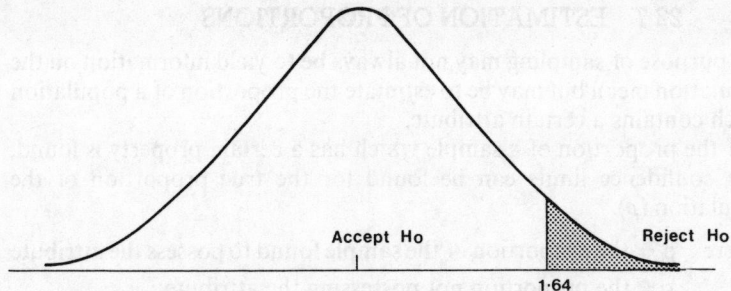

Fig. 22.3 One-tail significance test (0.05 level)

22.6 SMALL SAMPLES

If a sample of less than 30 is taken, then it is not possible to assume that the distribution of sample means is normal.

In this case, a theoretical sampling distribution called the t distribution can be used to enable confidence limits to be drawn up. This is also a symmetrical distribution, but the distribution depends upon the value (n − 1) where n is the size of the sample. The value (n − 1) is termed the number of degrees of freedom and a table of t values differs from the z value table in that the corresponding areas vary according to the number of degrees of freedom.

For instance, we can calculate a 95 per cent confidence interval for μ as:

$$\bar{x} \pm t_{0.025} \frac{s}{\sqrt{n}}$$

As can be seen from Fig. 22.4, $t_{0.025}$ replaces 1.96 and has a value which varies according to the number of degrees of freedom. Reference to a set of t distribution tables which shows the $t_{0.025}$, etc values with the appropriate number of degrees of freedom is required.

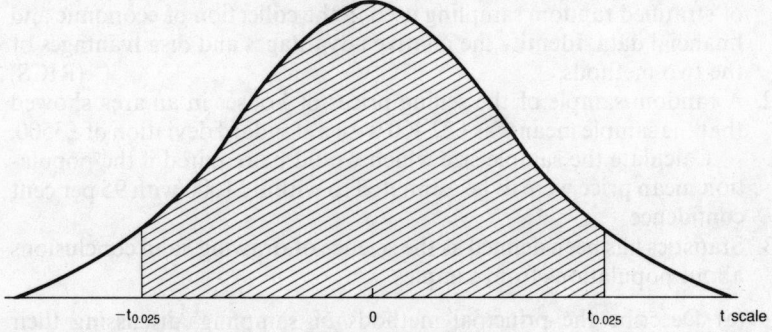

Fig. 22.4 t distribution with 95 per cent points

22.7 ESTIMATION OF PROPORTIONS

The purpose of sampling may not always be to yield information on the population mean but may be to estimate the proportion of a population which contains a certain attribute.

If the proportion of a sample which has a certain property is found, then confidence limits can be found for the true proportion of the population (ρ).

Where p = the proportion of the sample found to possess the attribute

q = the proportion not possessing the attribute

and n = the size of the sample

then a 95 per cent confidence interval for ρ is given by

$$\rho \pm 1.96 \sqrt{\frac{pq}{n}}$$

ie $\sqrt{(pq/n)}$ is the standard error.

Example 22.3

A random sample of 800 houses in a town showed that 480 of the dwellings had a central heating system. Estimate the proportion of houses in the whole town with central heating.

The 95 per cent confidence limits are therefore:

$$0.6 \pm 1.96 \sqrt{\left(\frac{0.6 \times 0.4}{800}\right)} = 0.6 \pm 0.0339$$

ie the population proportion is estimated to be between 0.5661 and 0.6339.

22.8 EXERCISES

1. Explain and illustrate the principles of simple random sampling and of stratified random sampling used in the collection of economic and financial data. Identify the relative advantages and disadvantages of the two methods. (RICS)
2. A random sample of the selling prices of houses in an area showed that the sample mean was £22 400 with a standard deviation of £3500.

    Calculate the sample size which would be required if the population mean price were to be estimated to within £1000 with 95 per cent confidence.
3. Statistics has been defined as the science of drawing valid conclusions about populations from samples.

    (a) Describe the principal methods of sampling, discussing their particular uses and limitations.

Exercises

(b) An estate agent wishes to estimate the typical cost of accommodation in his town. He takes a sample of eighty-one houses from his recent circulars. The prices of these houses show an arithmetic mean of £33 165 and a standard deviation of £6500.

On the assumption that the data represent a valid sample and this mean is also appropriate, estimate the limits of the population mean with 95 per cent confidence.

(c) What other reservations should we place on the assumption that £33 165 is a fair estimate of the cost of accommodation in his town?

III DATA PROCESSING SYSTEMS

23 COMPUTER SYSTEMS

23.1 INTRODUCTION

More and more use is being made of computers in the building industry, and even at technician level, a knowledge of at least the rudimentaries of computer systems is a necessary asset.

It cannot be the aim of this text to go into any depth on the technicalities of computer systems. All that is intended is an explanation of the functions of computer systems and their potential uses in the industry.

23.2 THE COMPONENTS OF A COMPUTER SYSTEM

Computers are of two main types–analog and digital–but for most work involving large amounts of calculations the latter type is the more important.

The digital computer performs calculations using numerical digits in a coded form such as the binary system (where every number is represented by a series of 0s or 1s) and this type of computer is ideally suited to statistical calculations and general data processing.

The physical units of a computer are termed hardware–as distinct from software, which refers to the programs which can be used on a particular computer system.

The basic units of a digital computer and the relationships between them are shown in Fig. 23.1.

The program control unit, arithmetic unit and storage are included in the central processor. Their individual functions are as follows:

1. The program control unit acts as the co-ordinator in the computer and its role is to see that instructions are followed. It also controls the

The components of a computer system

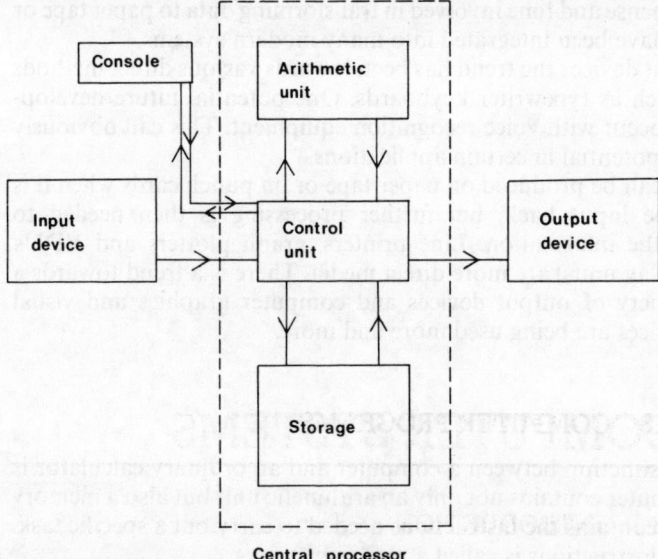

Fig. 23.1 Basic units of a computer system

transfer to and from the storage devices and to and from the input and output devices.
2. The arithmetic unit carries out the arithmetic operations and has a section where logical processing takes place.
3. Storage devices provide areas into which vast amounts of data can be fed. It is this type of unit which distinguishes the computer from an ordinary calculator. It stores the program of instructions as well as the data being used.

These components of the central processor are linked to the peripheral system of input and output devices. If such devices are linked directly to the computer they are termed on-line, whereas those not so linked are off-line.

Input devices enable numerical data and instructions to be fed into the computer in a form that can be identified by the machine. Input media take several forms depending on the application. Punch cards have been used since the earliest days of computers, as also has punched paper tape on which data are recorded by punching holes on a continuous strip of paper. Magnetic tape and magnetic discs as input media have become more popular and cheaper since the 1970s. Systems such as MICR (magnetic ink character recognition) and OCR (optical character recognition), which allow original documents to be read, can

Computer systems

save the expense and time involved in transforming data to paper tape or cards and have been integrated into many modern systems.

For input devices the trend has been towards various direct methods of entry such as typewriter keyboards. One potential future development may occur with voice recognition equipment. This can obviously have great potential in certain applications.

Output can be produced on paper tape or on punch cards when it is going to be input back, but further processing is then needed to transform the information. Line printers, graph plotters and VDUs (visual display units) are more direct media. There is a trend towards a greater variety of output devices and computer graphics and visual display devices are being used more and more.

23.3 COMPUTER PROGRAMS

The real distinction between a computer and an ordinary calculator is that a computer contains not only an arithmetic unit but also a memory unit which contains the instructions needed to carry out a specific task. This set of instructions is called a program.

A program has to be broken down into a series of simple instructions in numerical order and in logical sequence. This is termed **machine code** programming when the instructions are contained within the central processor.

The normal method of program coding is by the use of a program language. Such languages are a mixture of English and algebraic notation. A program called a compiler is stored in a computer so that when a language program is fed into the computer the computer translates into machine code.

It is obviously important that every computer does not have to 'understand' a different program language. Languages which are largely particular machine-oriented are usually termed low-level languages.

The following high-level languages are important as they are basically machine independent and are in widespread use:

1. ALGOL (ALGOrithmic Language), an international language, was developed as a scientific programming language.
2. BASIC is a very simple language and important in its use with microcomputers due to its simplicity for first-time users.
3. COBOL (COmmon Business Orientated Language) is a language for general commercial usage, such as payroll processing.
4. FORTRAN (FORmula TRANslating) is also a scientific language. It was the first high-level language and was developed by IBM. PL/1 (Programming Language 1) has also been developed by IBM with the aim of combining the features of a commercial and a scientific language, ie a language able to handle large volumes of data and able to deal with scientific routines.

23.4 FLOWCHARTS

The basic aim of a computer program is to present the computer with an ordered set of instructions which define the procedure required in a logical and precise manner.

A set of statements defining such a procedure is termed an algorithm.

A useful initial step in the formulation of an algorithm is a flowchart which is a diagram showing the main logical order of the stages in a program. The flowchart itself can develop from stage to stage in the process, eg outline, detail and program flowcharts, but only a basic flowchart can be considered here.

A conventional flowchart runs from top to bottom and consists of a set of boxes connected by flowlines and these boxes or symbols can be of several different types;

(a) process boxes representing some kind of operation, eg changing a value;
(b) input/output boxes representing an input/output function;
(c) decision boxes representing the decisions on which branches of a program to follow;
(d) termination symbols showing the beginning or end of a flowchart;
(e) connector symbols showing entry from or exit to another point in the same flowchart.

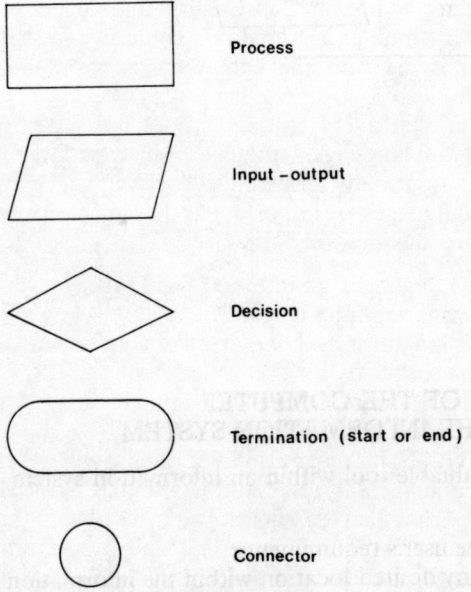

Fig. 23.2 Flowchart symbols

Computer systems

These symbols are depicted in Fig. 23.2. A simple example can illustrate the use of a flowchart.

Information on the examination marks of a set of first year students is recorded on a file. To progress to the second year a student must have obtained over 70 per cent in either his surveying or his statistics examination. A list is required of the successful students. A flowchart setting out logical steps to the solution of this problem is shown in Fig. 23.3.

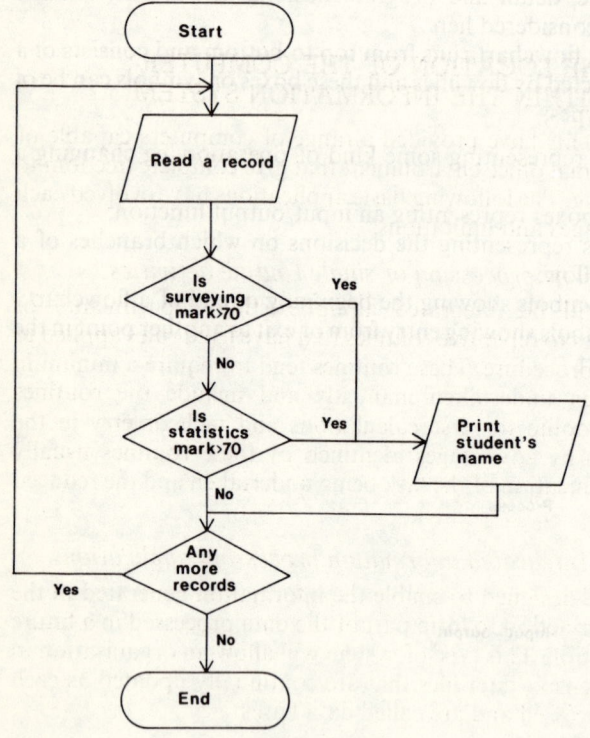

Fig. 23.3 A simple flowchart

23.5 THE ROLE OF THE COMPUTER WITHIN THE INFORMATION SYSTEM

Modern computers are a valuable tool within an information system, since they can:

(a) select data relevant to the user's requirements;
(b) provide data access at any desired location within the information system;

(c) enable information to be shared with a range of users;
(d) provide fast data processing capabilities;
(e) retrieve information in new combinations from the input format;
(f) provide an environment in which the mutual standards and goals of all the users of the information system can be identified and structured.

Only by understanding a user's requirements and selecting the correct computer hardware which incorporates software related to the application can the above characteristics be identified.

23.6 THE LOCATION OF THE COMPUTER WITHIN THE INFORMATION SYSTEM

Present developments have provided a range of computers capable of operating in a normal office environment that give complete freedom to the system planning. The following basic applications have evolved each having its advantages and limitations.

23.6.1 The processing of single routine activities

The computer will be integrated along existing departmental or organisational lines to repetitive routines that have fixed rules applied to each stage of the procedure. These routines tend to require a minimum of judgment when undertaken manually and include the routines outlined for theodolite traverse calculations and tacheometry in the earlier chapters. Any advantages identified by these routines usually relate to the simplification of the task being undertaken and the reduced time required.

23.6.2 Integrated information processing applications

These systems are designed to enable the information generated in the course of one transaction to form part of the data processed in a future computer application. This type of system will allow an organisation to develop comprehensive data files that are continually updated as each transaction is processed and are called data bases.

23.6.3 Data bases

Information systems of this type will require a high level of systems analysis at the implementation stage to identify the required characteristics of the data demanded at each level of the information system. This analysis will enable the relevant data files to be constructed and should minimise any duplication of data stored in file. Although data storage and reporting will be more efficient, a more complex code structure than that for single routine activities will be required and possibly routines to operate the system and provide data security will be required.

The high investment cost of this type of information system means that the user must identify more benefits than those identified for a single

Computer systems

linear application. Providing all the demands of the information system have been identified, the following benefits should be evident.

(a) accurate and rapid data capture at the source;
(b) immediate response to an enquiry;
(c) decisions based on timely and accurate data;
(d) computer power where required;
(e) a sharing of common information within the organisational structure.

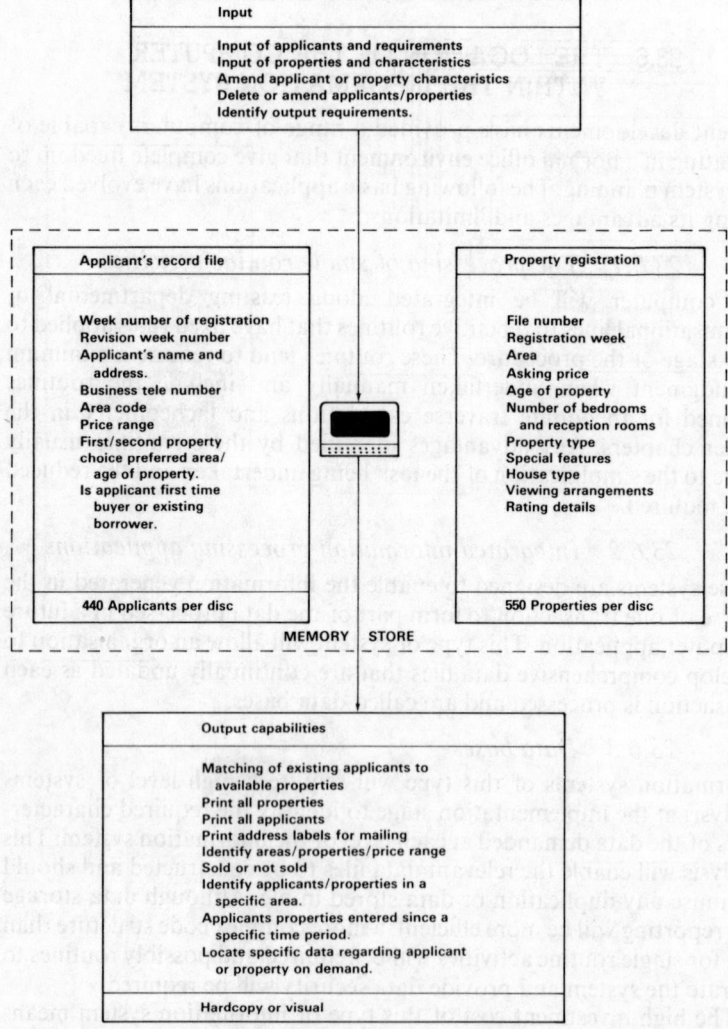

Fig. 23.4 Data sorting for property sales (courtesy Microsense Ltd)

The location of the computer within the information system

An example of a small data base routine presently operating on an Apple micro-computer is illustrated in Fig. 23.4. This shows data sorting for property sales where a number of reports or direct queries can be made regarding the status of properties and applicants. The package outlined is beneficial to a single or small group user who does not have the resources to develop sophisticated systems, since he can use the standard report formats provided by the package, or make minor adjustments to satisfy his organisation.

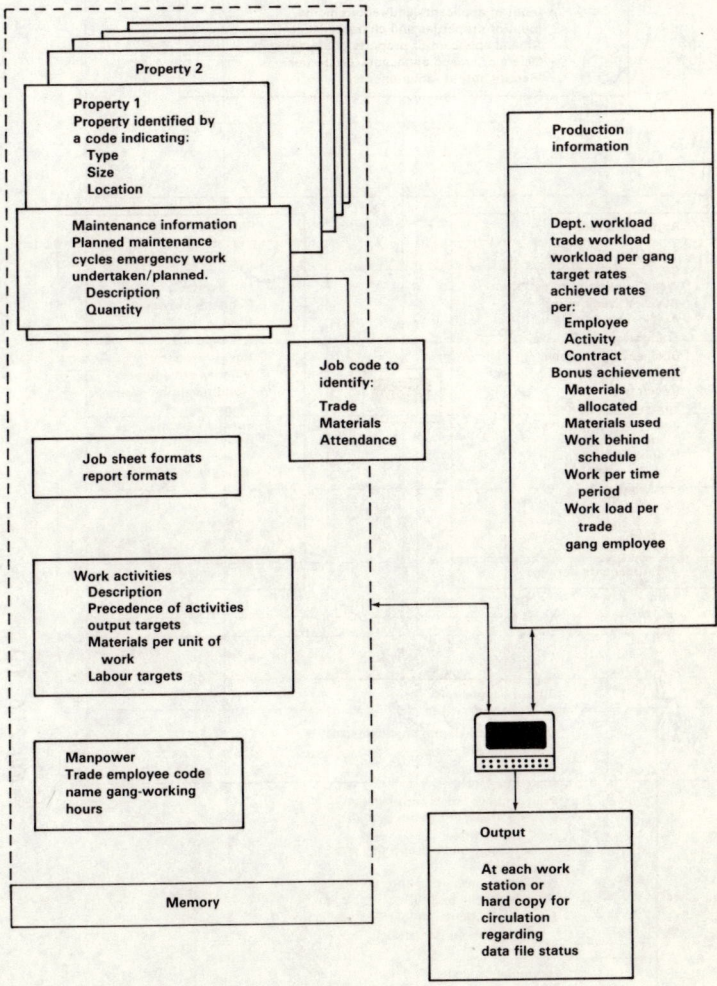

Fig. 23.5 Property maintenance

Computer systems

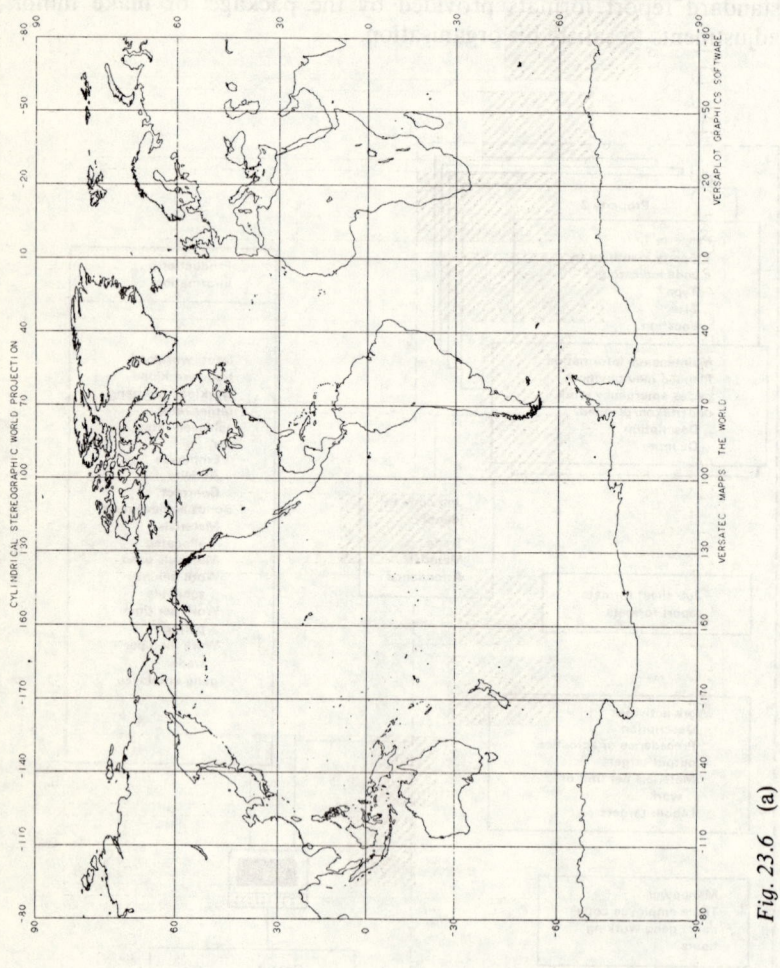

Fig. 23.6 (a)

The location of the computer within the information system

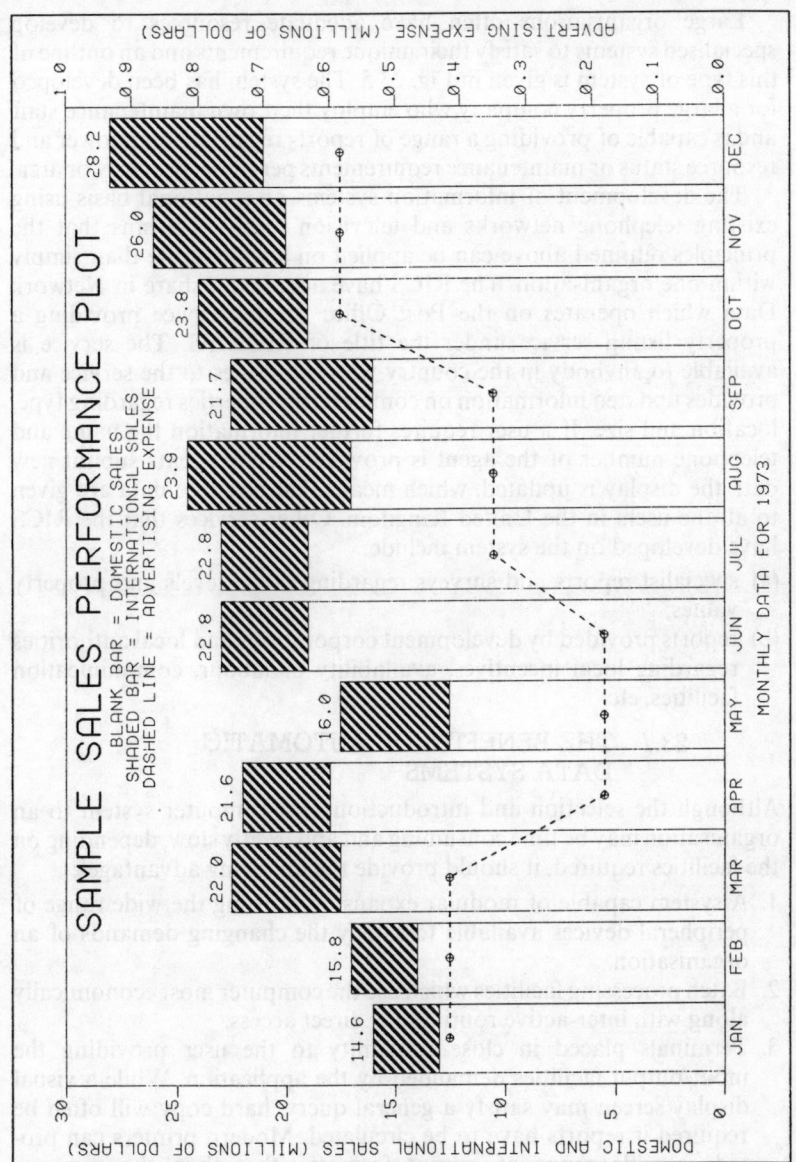

Fig. 23.6 (b) Output formats (courtesy Versatec)

Large organisations often have adequate resources to develop specialised systems to satisfy their unique requirements and an outline of this type of system is given in Fig. 23.5. The system has been developed for a large property company who employ their own maintenance staff and is capable of providing a range of reports regarding manpower and resource status or maintenance requirements per trade, property or area.

The development of information systems on a national basis using existing telephone networks and television receivers means that the principles outlined above can be applied on a wider scale than simply within one organisation. The RICS have acquired a share in Network Data which operates on the Post Office Prestel service providing a property listing service under the title of RICS/EG. The service is available to anybody in the country who subscribes to the service and provides updated information on commercial properties regarding type, location and size. If a user requires further information the name and telephone number of the agent is provided. When agents submit new data the display is updated, which means that accurate data are given to all the users in the United Kingdom. Other services that the RICS have developed on the system include:

(a) specialist reports and surveys regarding rental levels and property values;
(b) reports provided by development corporations and local authorities regarding local incentives, availability of labour, communication facilities, etc.

23.7 THE BENEFITS OF AUTOMATIC DATA SYSTEMS

Although the selection and introduction of a computer system to an organisation may be time consuming and apparently slow, depending on the facilities required, it should provide the following advantages:

1. A system capable of modular expansion utilising the wide range of peripheral devices available to satisfy the changing demands of an organisation.
2. Batch processing facilities which use the computer most economically along with inter-active routines for direct access.
3. Terminals placed in close proximity to the user providing the input/output facilities demanded by the application. While a visual display screen may satisfy a general query, hard copy will often be required if reports have to be circulated. Modern printers can provide a wide range of output formats other than the common alpha/numeric formats that have been used in the past. Examples of these are shown in Fig. 23.6.
4. Automatic data sorting by the central processor will reduce manual demands.
5. Data filing, storing and sorting procedures are undertaken by the software.

23.8 EXERCISES

1. Explain the difference between computer software and hardware, giving an example of the former and the essential parts of the latter.
(RICS)

2. A materials supplier owns a series of retail outlets and pays each manager a monthly bonus according to his shop's turnover for that month. If the turnover is less than £20 000 the bonus paid is zero; if the turnover is £20 000 and less than £30 000 the bonus paid is £100, and if the turnover is £30 000 or above a bonus of £200 is paid. If the name of each manager is recorded with his shop's monthly turnover on file, draw a flowchart which would enable the name of each manager together with his bonus to be printed.

3. The availability of inexpensive micro-computers has enabled even small practices and businesses to make use of computing facilities. In a general practice:
   (a) to what applications might a micro-computer be put?
   (b) In which of these, if any, might 'software' be a problem?
   (c) To what extent might a 'mainframe computer' (a full-size machine) still be preferred?
(RICS)

APPENDIX

THE NORMAL DISTRIBUTION FUNCTION

Z	Area		Z	Area		Z	Area	
0.00	0.5000	40	0.25	0.5987	39	0.50	0.6915	35
0.01	0.5040	40	0.26	0.6026	38	0.51	0.6950	35
0.02	0.5080	40	0.27	0.6064	39	0.52	0.6985	34
0.03	0.5120	40	0.28	0.6103	38	0.53	0.7019	35
0.04	0.5160	39	0.29	0.6141	38	0.54	0.7054	34
0.05	0.5199	40	0.30	0.6179	38	0.55	0.7088	35
0.06	0.5239	40	0.31	0.6217	38	0.56	0.7123	34
0.07	0.5279	40	0.32	0.6255	38	0.57	0.7157	33
0.08	0.5319	40	0.33	0.6293	38	0.58	0.7190	34
0.09	0.5359	39	0.34	0.6331	37	0.59	0.7224	33
0.10	0.5398	40	0.35	0.6368	38	0.60	0.7257	34
0.11	0.5438	40	0.36	0.6406	37	0.61	0.7291	33
0.12	0.5478	39	0.37	0.6443	37	0.62	0.7324	33
0.13	0.5517	40	0.38	0.6480	37	0.63	0.7357	32
0.14	0.5557	39	0.39	0.6517	37	0.64	0.7389	33
0.15	0.5596	40	0.40	0.6554	37	0.65	0.7422	32
0.16	0.5636	39	0.41	0.6591	37	0.66	0.7454	32
0.17	0.5675	39	0.42	0.6628	36	0.67	0.7486	31
0.18	0.5714	39	0.43	0.6664	36	0.68	0.7517	18
0.19	0.5753	40	0.44	0.6700	36	0.69	0.7549	31
0.20	0.5793	39	0.45	0.6736	36	0.70	0.7580	31
0.21	0.5832	39	0.46	0.6772	36	0.71	0.7611	31
0.22	0.5871	39	0.47	0.6808	36	0.72	0.7642	31
0.23	0.5910	38	0.48	0.6844	35	0.73	0.7673	31
0.24	0.5948	39	0.49	0.6879	36	0.74	0.7704	30

THE NORMAL DISTRIBUTION FUNCTION

Z	Area		Z	Area		Z	Area	
0.75	0.7734	30	1.15	0.8749	21	1.55	0.9394	12
0.76	0.7764	30	1.16	0.8770	20	1.56	0.9406	12
0.77	0.7794	29	1.17	0.8790	20	1.57	0.9418	11
0.78	0.7823	29	1.18	0.8810	20	0.58	0.9429	12
0.79	0.7852	29	1.19	0.8830	19	1.59	0.9441	11
0.80	0.7881	29	1.20	0.8849	20	1.60	0.9452	11
0.81	0.7910	29	1.21	0.8869	19	1.61	0.9463	11
0.82	0.7939	28	1.22	0.8888	19	1.62	0.9474	10
0.83	0.7967	28	1.23	0.8907	18	1.63	0.9484	11
0.84	0.7995	28	1.24	0.8925	19	1.64	0.9495	10
0.85	0.8023	28	1.25	0.8944	18	1.65	0.9505	10
0.86	0.8051	27	1.26	0.8962	18	1.66	0.9515	10
0.87	0.8078	28	1.27	0.8980	17	1.67	0.9525	10
0.88	0.8106	27	1.28	0.8997	18	1.68	0.9535	10
0.89	0.8133	26	1.29	0.9015	17	1.69	0.9545	9
0.90	0.8159	27	1.30	0.9032	17	1.70	0.9554	10
0.91	0.8186	26	1.31	0.9049	17	1.71	0.9564	9
0.92	0.8212	26	1.32	0.9066	16	1.72	0.9573	9
0.93	0.8238	26	1.33	0.9082	17	1.73	0.9582	9
0.94	0.8264	25	1.34	0.9099	16	1.74	0.9591	8
0.95	0.8289	26	1.35	0.9115	16	1.75	0.9599	9
0.96	0.8315	25	1.36	0.9131	16	1.76	0.9608	8
0.97	0.8340	25	1.37	0.9147	15	1.77	0.9616	9
0.98	0.8365	24	1.38	0.9162	15	1.78	0.9625	8
0.99	0.8389	24	1.39	0.9177	15	1.79	0.9633	8
1.00	0.8413	25	1.40	0.9192	15	1.80	0.9641	8
1.01	0.8438	23	1.41	0.9207	15	1.81	0.9649	7
1.02	0.8461	24	1.42	0.9222	14	1.82	0.9656	8
1.03	0.8485	23	1.43	0.9236	15	1.83	0.9664	7
1.04	0.8508	23	1.44	0.9251	14	1.84	0.9671	7
1.05	0.8531	23	1.45	0.9265	14	1.85	0.9678	8
1.06	0.8554	23	1.46	0.9279	13	1.86	0.9686	7
1.07	0.8577	22	1.47	0.9292	14	1.87	0.9693	6
1.08	0.8599	22	1.48	0.9306	13	1.88	0.9699	7
1.09	0.8621	22	1.49	0.9319	13	1.89	0.9706	7
1.10	0.8643	22	1.50	0.9332	13	1.90	0.9713	6
1.11	0.8665	21	1.51	0.9345	12	1.91	0.9719	7
1.12	0.8686	22	1.52	0.9357	13	1.92	0.9726	6
1.13	0.8708	21	1.53	0.9370	12	1.93	0.9732	6
1.14	0.8729	20	1.54	0.9382	12	1.94	0.9738	6

THE NORMAL DISTRIBUTION FUNCTION

Z	Area		Z	Area		Z	Area	
1.95	0.9744		2.35	0.99061		2.75	0.99702	
1.96	0.9750	6	2.36	0.99086	25	2.76	0.99711	9
1.97	0.9756	6	2.37	0.99111	25	2.77	0.99720	9
1.98	0.9761	5	2.38	0.99134	23	2.78	0.99728	8
1.99	0.9767	6	2.39	0.99158	24	2.79	0.99736	8
		5			22			8
2.00	0.97725		2.40	0.99180		2.80	0.99744	
2.01	0.97778	53	2.41	0.99202	22	2.81	0.99752	8
2.02	0.97831	53	2.42	0.99224	22	2.82	0.99760	8
2.03	0.97882	51	2.43	0.99245	21	2.83	0.99767	7
2.04	0.97932	50	2.44	0.99266	21	2.84	0.99774	7
		50			20			7
2.05	0.97982		2.45	0.99286		2.85	0.99781	
2.06	0.98030	48	2.46	0.99305	19	2.86	0.99788	7
2.07	0.98077	47	2.47	0.99324	19	2.87	0.99795	7
2.08	0.98124	47	2.48	0.99343	19	2.88	0.99801	6
2.09	0.98169	45	2.49	0.99361	18	2.89	0.99807	6
		45			18			6
2.10	0.98214		2.50	0.99379		2.90	0.99813	
2.11	0.98257	43	2.51	0.99396	17	2.91	0.99819	6
2.12	0.98300	43	2.52	0.99413	17	2.92	0.99825	6
2.13	0.98341	41	2.53	0.99430	17	2.93	0.99831	6
2.14	0.98382	41	2.54	0.99446	16	2.94	0.99836	5
		40			15			5
2.15	0.98422		2.55	0.99461		2.95	0.99841	
2.16	0.98461	39	2.56	0.99477	16	2.96	0.99846	5
2.17	0.98500	39	2.57	0.99492	15	2.97	0.99851	5
2.18	0.98537	37	2.58	0.99506	14	2.98	0.99856	5
2.19	0.98574	37	2.59	0.99520	14	2.99	0.99861	5
		36			14			4
2.20	0.98610		2.60	0.99534		3.0	0.99865	
2.21	0.98645	35	2.61	0.99547	13	3.1	0.99903	38
2.22	0.98679	34	2.62	0.99560	13	3.2	0.99931	28
2.23	0.98713	34	2.63	0.99573	13	3.3	0.99952	21
2.24	0.98745	32	2.64	0.99585	12	3.4	0.99966	14
		33			13			11
2.25	0.98778		2.65	0.99598		3.5	0.99977	
2.26	0.98809	31	2.66	0.99609	11	3.6	0.99984	7
2.27	0.98840	31	2.67	0.99621	12	3.7	0.99989	5
2.28	0.98870	30	2.68	0.99632	11	3.8	0.99993	4
2.29	0.98899	29	2.69	0.99643	11	3.9	0.99995	2
		29			10			2
2.30	0.98928		2.70	0.99653				
2.31	0.98956	28	2.71	0.99664	11			
2.32	0.98983	27	2.72	0.99674	10			
2.33	0.99010	27	2.73	0.99683	9			
2.34	0.99036	26	2.74	0.99693	10			
		25			9			

Source: Cambridge Elementary Statistical Tables, Lindley and Miller, Cambridge University Press.

ANSWERS TO SELECTED EXERCISES

Chapter 3
Triangle ABC error = 0.4 per cent; triangle XYZ error = 9.26 per cent.

Chapter 4
Ex. 1 True forward bearing is 329°

Chapter 6
Ex. 1 Distance AE is 914.442 m with a bearing of S 19°23'E from Station A.
Ex. 2 Error to be corrected in partial co-ordinates is E -0.112 m and N $+0.047$ m.

Chapter 7
Ex. 3 Collimation error 1 in 7.037 m and true difference in level is 2.543 m.

Chapter 8
Ex. 1 Reduced level on temporary bench mark is 21.528 m.
Ex. 3 Reduced level at proposed manhole is 103.459 m.

Chapter 9
Ex. 1 Difference in level is 17.35 m.
Ex. 2 Gradient between A and C is 1 in 7.02.

Chapter 11
Ex. 1 Area between chain line and boundary is 73.4 m^2.
Ex. 2 Volume of excavation is 104.95 m^3.
Ex. 3 Area of land is (a) 574.67 m^2; (b) 570 m^2; (c) 636 m^2.

Chapter 12
Ex. 1 Invert level of manhole is 47.55 m.

Answers to selected exercises

Chapter 15
Ex. 1 Arithmetic mean = $4.85 \, m^2$.
Median (by calculation) = $4.844 \, m^2$.
Mode (by calculation) = $4.833 \, m^2$.
Ex. 3 Arithmetic mean = £52.4.

Chapter 16
Ex. 1 Building technology. Mean = 54.5 marks.
Standard deviation = 18 marks.
Construction. Mean = 54.875 marks.
Standard deviation = 13.54 marks.
Ex. 2 Median = £28 855.2.
Quartile deviation = £2071.3.
Ex. 3 Arithmetic mean = 9.62 km.
Standard deviation = 3.57 km.

Chapter 17
Ex. 1 (a) 0.0005; (b) 0.0440; (c) 0.0006; (d) 0.9506
Ex. 2 (a) P(machine 2 idle) = 0.1
P(machine 3 working) = 0.5;
(b) 0.210; (c) 0.648; (d) 0.166
Ex. 3 (a) P(teletype) = 0.68;
(b) P(both use same type of machine) = 0.559

Chapter 18
Ex. 1 (b) (i) If an aggregative base weighted price index is compiled using 1977 as base year and taking 1977 quantities as weights, the index series is:
1977   100
1978   116.4
1979   149.1
(ii) If an aggregative base weighted quantity index is compiled using 1977 as base year and taking 1977 prices as weights, the index series is:
1977   100
1978   113.6
1979   138.2
Ex. 2 Using 1969 prices as weights, the index series is:
1969   100
1970   118.3
1971   139.1
1972   169.6
Ex. 3 (c) 1978   100
1979   120.1
1980   143.3

350

Answers to selected exercises

Chapter 19
Ex. 2 Y = 4.2985 + 0.0619 X
   (i) £7.39(35)
   (ii) £10.48(85)
Ex. 3 Estimated population (000's)
   1941 = 15.7
   2001 = 29.4

Chapter 20
Ex. 1 r = +0.790
Ex. 3 r = +0.919

Chapter 21
Ex. 1 Using a three year moving average, the trend is calculated as:

Year	Trend
1965	109
1966	113
1967	118
1968	121
1969	125
1970	128
1971	132

Ex. 2 The average seasonal variations are:

1	11	111	1V
+99	−43.5	−32	−23

The estimates for the last three quarters of 1974 are:

Quarter	11	111	1V
£(m)	57(.6875)	72(.75)	85(.3125)

(These estimates are based on the addition of the 1974 (1) value to the time series.)

Chapter 22
Ex. 2 Minimum sample size of 48
Ex. 3 (b) 95 per cent confidence limits: £31 749 to £34 581.

INDEX

Abney level 36
Accuracy, general 2–4
  for specific applications see
    technique
Accromatic lens 161
Addition rule (of probability) 265
Additive constant 140
Aggregative index 282–4
Air photogrammetry 160
  advantages 160
  application 160
  characteristics of 170
  determination of ground height
    from 169
  distortion of 163, 166
  equipment for 161
  plotting from 168
  principles of 160
  rectified 216
Algorithm 337
Alternative hypothesis 328
Anallatic lens 141
Angles
  booking of 69
  depression 55
  elevation 54
  horizontal 54
  measurement of 66, 70
  rounds of 68
  vertical 70
Arbitrary meridian 46, 77

Areas
  defined by
    co-ordinates 183
    graphical methods 174
    latitudes and departures 183
    mid-ordinate rule 175
    planimeter 181
    Simpson's rule 179
    straight lines 173
    trapezoidal rule 177
Arithmetic mean 240–4, 249, 260
  group data 242
  ungrouped data 241
  weighted 243
Arrows 26
Assumed mean method 243
Atmospheric refraction 128
Automatic level 106
Auxiliary base 157
Averages Ch. 15
Azimuth 54

Backsight 43, 49, 77, 120, 184
Ball and socket mounting 106, 107
Band steel 26, 27, 28, 77
Bar charts 234–5
  component 235
  percentage 236
Base line 3, 24, 25, 204
  auxiliary base 157

Index

chain 26, 27, 28
electronic distance measurement of 78
setting out from 201
vertical air photogrammetry 161
Base period 285–6
Basic survey procedures 3
Batter 200
Beam compass 41
Bearing 43
  arbitrary 46
  back bearing 43
  forward 43
  magnetic 45
  quadrantal 44
  reduced 44, 80, 81, 90
  true 45
  whole circle 43
Bench marks
  ordnance survey 118
  temporary 119
Bias 322
Binomial distribution 268–9
Block plan 198
Booking procedures
  building survey 217
  chain survey 33
  levels 121
  tacheometry 148
  theodolite 67, 68, 69
Bolt bench mark 119
Bowditch's method of adjustment
  calculation 92
  graphical 50
British standards 26
Bubble
  accuracy 108
  adjustment 105, 113
  arrangement 103, 108
Building research station 40, 62, 74
Building survey 212
  booking procedures for 213
  equipment for 212
  rectified photogrammetry 216
  procedures for 213
Bureaux facilities (computer)
  tacheometry 152
  traverse survey 96

Cadastral survey 3, 75
Calculators electronic 95, 150
Cameras
  air survey 161
  rectified photogrammetry 217
  stereoscopy 167
Central limit theorem 325
Central tendency, measures of 240
Chain
  accuracy 27
  description 26, 27, 28
Chain-base index numbers 286
Chain survey
  accuracy of 39
  booking procedures for 33
  control of 28
  equipment for 26
  error correction of 31, 35, 40, 41
  location of detail 31
  obstructions to 37
  plotting of 41
  principles of 24
Chaining a line 29
Chaining on a slope 35
Change plate, levelling 112
Change point 120, 150
Check lines 26, 76
Classification 224
Class interval 224–5
  limits 225
  mark 233
Clinometer 35
Clip screw 55, 66
Closed traverse 76
Closing error 50
  compass traverse 48
  theodolite 79
Coefficient of variation 259
Coincidence bubble 103
Collimation error 112
Collimation, height of 123
Collimation line 118, 123
Combinations 267
Compass bearings 45
Compass prismatic 46
Compass traverse
  adjustment of 50
  correction of readings for 49, 50
  definition of 48
  plotting 49
  procedure for 48
Component bar charts 235–6
Computer applications 339–44

353

Index

Computer programs 336
Computer services
 tacheometry 152
 traverse survey 96
Computer systems Ch. 23
Conditional probability 266
Confidence limits 326–30
Conformal maps 8
Conical projections 11
Continuous variables 225
Contour lines
 accuracy of 130
 definition of 129
 volume calculation from 193
Contouring
 direct method 130
 grid method 130
Convergence, National Grid 13
Control
 points 3, 22, 23, 160
 principles of 3, 4, 26, 28
 Tension 77
Co-ordinates 49, 76, 120, 140, 199, 201
 adjustment of 91
 areas from 183
 National Grid 14
 partial 80
 rectangular traverse 81, 91, 94
 setting out from 206
Corrections to
 chain line 40
 co-ordinates 91
 rectilinear area 177
 slope distance 35
 temperature change 40
Correlation Ch. 20
Critical values 328
Cross hairs 56, 104, 139
Cross-sections 187
Cumulative frequency distribution 226–7
Curvature of earth 117, 128
Cuttings 187, 200
Cyclical variations 312–3

Data 222–4
 continuous 225, 232
 discrete 225
Data bases 339–44

Datum line
 building survey 216
 levelling
  arbitrary 102, 117
  ordnance survey 22, 117
Deciles 249–50
Declination magnetic 45
Degrees of freedom 329
Departure 81, 91, 184
Dependent variable 293
Determination of ground height 162, 163
Deviations using a theodolite 62, 71, 74
Diaphragm 55, 58, 104, 139
Direction correction 13
Dispersion Ch. 16
Distance wedge optical 157
Diurnal variation 45
Distortion of aerial photography 163
Double reading optical micrometer 61
Drains setting out 198
Drop arrow 26
Dumpy level
 accuracy of 107, 108
 adjustment of 112, 165
 description of 104
 use of 7, 109

Earth curvature 2, 9, 117, 128
Eccentricity, errors from 64, 68, 137, 142, 146, 161
Electro-magnetic distance measurement 4, 78
Embankments, setting out 199
End areas 186, 188
Engineering survey 3, 80, 126
Engineer's
 level 105, 107, 108
 theodolite 56, 63
Equal area map projections 8
Errors
 accidental 5, 39
 angular measurement 30, 71
 classification of 5, 39, 71, 79, 80, 163
 horizontal staff tacheometry 155
 levelling 126

354

Index

linear measurement 30, 35, 40, 77, 91
magnetic bearing 45, 50
ordnance survey maps 21, 23
vertical staff tacheometry 135, 140, 142, 146
Estimation
  of population mean 326–7
  of population proportions 330
Extrapolation 294, 315–17, 320
Eye base stereoscopy 167
Eyepiece 58, 106, 140

Face, change of 68
Factorial notation 267
Fibreglass tape 27
Fiducial marks 161
Fieldnotes
  building survey 213
  chain survey 34
  general 6
  levelling 121
  tacheometry 147
  traverse 77, 86, 98
Flow charts 337–8
Flush brackets 112
Focal length 57, 63, 108
Focussing, telescope 57
Follower 29
Footscrew 55, 64, 105
Foresight 120
Formation level 198
Framework control of 4, 22
Frequency
  cumulative 226–7
  curve 232–3
  distribution 224–8
  polygon 231–2, 260
  table 222–4, 231

Geodetic survey 2
Geometric mean 248
Give and take lines 175
Glass circle theodolite 60
Gnomonic projections, maps 10
Gradient 35, 130, 145, 188, 198
Graphical adjustment of a traverse 50
Graphical presentation 228–37
Graticule line
  see also cross-hairs 56, 138

Great circle 10
Grid
  contour method of 130
  co-ordinates 13, 80, 87
  distance 16
  national 13
  setting out 207
  volume calculation from 194
Grid north 14, 46
Ground control, aerial survey 160
Ground principal point 161

Hand level 36
Hardware 334, 339
Harmonic mean 248
Hectare 173
Height reading curve 154
Histogram 229–32
  with unequal class interval 230
Horizontal
  angle 53, 66, 147
  axis 55
  equivalent 130
  line 102
Horizontal staff tacheometry
  optical wedge 157
  sub tense bar 155
Hypothesis testing 327–9

Inclined sights
  tacheometry 137, 142, 146
  theodolite 70, 72, 208
Independent
  events 265–6
  variable 293, 318
Index numbers Ch. 18
Instrument errors 39, 45, 71, 127
Intermediate sights 120, 149
Internal focusing telescope 56, 139
Interpolation 294
  method for contours 130
Interpretation of air photographs 170
Invar bands 27, 41, 77, 202
Invert level 198
Inverted staff reading 125
Isocentre 166

Key points 205

355

Index

Land surveying
  classes of 2, 24, 75, 107, 117, 135, 160
  definition of 2, 117
  principles of 3, 24, 48, 102, 136, 138, 155, 161
  types of error in 5, 21, 35, 39, 50, 79, 111, 126, 146, 163, 166
Laspeyres index 282–4
Lateral overlap 167
Latitude 80, 91
Lens
  camera arrangement of 161
  telescope arrangement of 56
Level book 121
Level of detail, ordnance survey 21
Level line 118
Levelling
  accuracy of instruments for 107
  adjustment of instruments for 109, 112
  booking procedures for 121
  definition of 118
  reciprocal 127
  reduction of 121
  staff 110
Line of collimation 55, 56, 119, 123
Line of best fit 292–5
Line pegs 198
Linear measurement
  accuracy of 22, 27, 39, 75, 135, 204
  check of 5, 78, 202
  equipment for 26, 78, 136, 138, 212
  errors in 5, 30, 35, 39, 41, 91, 135, 146
  slope correction of 35
Linear regression Ch. 19
Linear relationship 289–90
Local attraction, magnetic 45
Longitude 11, 45

Magnetic
  bearings 45
  compass 46
Magnetic declination 45
Maps
  accuracy of 21
  direction correction for 13
  grid referencing of 14
  ordnance survey 16, 117, 118
  projection systems for 8
  revision of 17
  scale factors 13
Mean
  deviation 253–5
  reading, angular 61
  sea level 117
Measure of location Ch. 15
Measurement
  angular 4, 31, 35, 46, 53, 75, 209
  linear, 4, 28, 75, 78, 135, 201
  obstacles to 37, 45, 135
  plus 30
  running 30, 213
Median 244–6
Mercator projection 12
Meridian
  arbitrary 46, 76
  grid 13, 46, 76
  secular 45
  true 45
Method of
  least squares 292–4, 317–20
  moving averages 311–17
Micro processor
  tacheometry 151
  theodolite 96
Micrometer optical 60
Mid-ordinate rule, areas 175
Mid-point 242
Mode 247–8, 249
Mosaics 170
Multiplication rule (of probability) 265
Multi-state sampling 324
Multiplying constant, tacheometry 139
Mutually exclusive events 265

Nadir point 166
National grid 14
Normal
  curve 269–77
  distribution 269–77
  equations 292–4, 318
North
  grid 14, 46, 77
  magnetic 45, 77
  true 45

356

Index

Northing 80, 91, 206
Null hypothesis 327–8
Obstacles to field measurement 37, 45, 125, 138, 142
Objects lens 57, 58
Observations, booking of
  angular 49, 67, 98
  chain survey 33, 214
  levelling 121, 125, 195
  principles 6
  tacheometry 148, 153
  theodolite 67, 98, 185
Offset pegs 198, 200
Offsets 30, 78, 177
Ogive 233–4, 246
One-tail significance test 328
Open end classes 225–6
Open traverse 77
Optical
  reading systems 61
  square 32, 203
  systems theodolite 60
  wedge 157
Optical distance measurement 152, 157, 169
Optical micrometer reading systems 60
Optical plummet 65, 209
Order of accuracy 2, 4, 8, 21, 30, 39, 50, 61, 75, 79, 107, 127, 146, 155, 163, 166
Ordnance survey
  accuracy of 21
  benchmarks for 118
  grid referencing 13, 23
  levelling 118
  maps 16
  sheets numbering 16
Orthographic projections 9
Orthomorphic projections 8, 9
Overlap, aerial photography 167

Paasche index 282–4
Parallactic angle 135
Parallax
  bar 169
  error 58
Parcel number 20
Pearson's
  coefficient of skewness 260–2
  correlation coefficient 300–3

Percentage bar chart 235–6
Percentiles 249–50
Pegs 26, 77, 113, 198, 200, 201, 203
Permanent adjustment of level
  automatic 114
  dumpy 113
  tilting 114
  theodolite 72
Permissible error in
  chain survey 39
  compass traverse 50
  levelling 126
  tacheometry 135, 146, 155
Permutations 267
Perspective centre 161
  projections 9
Photogrammetry *see* Air photogrammetry
Photographs, air
  characteristics of 170
  ground height from 169
  plotting maps from 168
  stereoscopy 167
Pie chart 237
Planimeter 181
Plumb bob 36, 64, 105, 208
Plus measurements 30
Polar co-ordinates 4, 50, 80
Precise
  levelling 22, 107
  traverse 2, 75
Price relative index 284–5
Principal point 161
Prismoidal rule 189, 193
Probability Ch. 17
  conditional 266
  rules of 265–6
Product moment correlation coefficient 299–303
Profile board 199, 202
Projections, maps 8
Prismatic compass 46
Programmable
  calculators
  tacheometry 151
  theodolite 95

Quadrant bearing 44, 80, 90
Quartile deviation 253, 262
Quartiles 249–50, 253

357

Index

Radial line plotting 168
Random number tables 323
Random sampling 322–3
Ranging a line 29
Ranging rod 26
Rank correlation 304–5
Rectangular co-ordinates 80, 91, 206
Rectangular grid 206
Rectified photogrammetry 216
Rectilinear area 176, 183
Reciprocal levelling 127
Reduced
  bearing 44, 80, 90
  level 102, 121
Reference
  pegs 205
  grid 194, 206
Regression line 293–5, 297, 318
Reiteration, method of 67
Relative frequency 264
Repetition angular measurement by 67
Representative fraction 163
Retail price index 285, 286
Reticule 56
Reversal 68, 104
Rhumb line 12
Right angle, setting out of 31, 202
Rise and fall method of booking 122
Running offsets 30

Sample 322–4
Sample proportions 330
Sample size 324
Sampling distributions 324–5
Sampling methods 322–4
  multi-stage 324
  stratified 323
  systematic 323
Scatter diagram 291–2, 297–9
Seasonal variations 311–12
Sections
  cross 187
  longitudinal 133
Semi-averages, method of 317
Sensitivity of bubble 65, 103, 108
Setting out
  buildings 201
  cuttings 200
  drains 198

  embankments 200
  grid 204
Sight rails 198, 203
  square 202
Significance testing 327–9
Simpson's rule
  areas 179
  volumes 189
Slope correction 35
  rail 200
Slow motion screw 54, 66, 109
Small samples 329
Software (computer) 96, 151, 334, 339
Spearman's coefficient of correlation 304–5
Spurious correlation 306
Standard error
  of mean 325–7, 330
  of proportions 330
Standard distribution 271–7
Station pegs 26, 77, 113
Steel tapes 26, 77, 202, 212
Stereographic projections 9
Stereoscopic pairs 167
Stratified sampling 323
Sub-tense bar 155
Survey
  aerial 160
  chain 24
  compass 48
  tacheometric 135
  theodolite 76

Tabulation 223–6
Tacheometry
  fixed hair 138
  moveable hair 136
  optical wedge 157
  sub-tense bar 155
  vertical staff 136
Tacheometer 152
Tacheometric staff 152
Tally marks 223
Tangent screw 54, 66
Tapes
  glass fibre 26
  linen 26
  steel 26, 77, 113
t-distribution 327–9
Telescope, internal focussing 56

Index

Temporary adjustment
  level 105
  theodolite 62
Temporary bench mark 119, 200, 201
Tension control 77
Tests of significance 327–9
Theodolite
  accuracy of 62
  axes of 55
  booking of observations 69
  description 53
  focussing 56
  permanent adjustment of 72
  reading systems for 59
  setting up the 58, 60, 62
  specification of 61
  traverse 75
  types of 58, 60
Tilting level
  description of 106
  permanent adjustment of 106
  temporary adjustment of 114
Tilting screw 106
Time series Ch. 21
Topographical survey 3
Transverse mercator projection 12, 14
Trapezium, area of 174, 183
Trapezoidal rule 177, 187
Traverse survey
  adjustment of 50, 92
  compass 48
  theodolite 75
Traveller 198

Trench, excavation control 199
Trend 311–17
Triangle, area of 174
Triangulation survey 4, 75, 79
Tribrach 55
Tripod 55, 62, 105
Trunnion axis 56
Two-peg test 113
Two-tail significance tests 328

Upper plate 54, 55, 59, 60, 61, 66

Variable 222
Variance 255–6
Vernier theodolite 59
Vertical angular measurement 70
  axis 54, 64, 105
  control 208, 210
  circle 54, 74, 208
Volume of earthworks
  contours 193
  cross-sections 187
  cut and fill 191
  method of end areas 186, 188
  prismoidal rule 190
  spot levels 194

Watkin's clinometer 36
Weighted mean 243–4
Whole-circle bearing 43, 87, 95
Wide-angle lens 161

Z scale (standard scores) 271–7, 325–6
Zero curve 154